ANALYSIS OF SHIP SPEED

AND ENGINE PARAMETERS IN THE TROPICS

ANATOLY ROZENBLAT

ISBN: 1-4033-8171-2 (e-book)
ISBN: 1-4033-8172-0 (Paperback)

This book is printed on acid free paper.

1stBooks – rev. 07/09/03

PREFACE TO THE FIRST EDITION

The motion of a ship through seawater in the tropics is made possible by a highly complex combination of factors affecting speed, efficiency, and performance. Of these factors, the temperature of the seawater surrounding the ship may hold the greatest impact on performance parameters.

Through five-in-depth papers included in *Collected Works of A.I.Rozenblat: Analysis of Ship Speed and Engine Parameters in the Tropics,* Anatoly I.Rozenblat explores the direct impact a variety of variable conditions have on the efficiency and performance of a ship. He arrives at this conclusions through detailed statistical analysis of such factors as exhaust temperature, water temperature, in-service time, wave action, and marine fouling on the ship's body.

Rozenblat supports his papers with numerous figures, graphs, and equations which describe and illuminate the complexities of his analysis. His careful attention to minute detail, technical subject, and knowledgeable discussion combine to make this a valuable and informative reference book for those interested in marine technology ,ship mechanics, and science in general.

Anatoly Rozenblat

CONTENTS

CHAPTER 1

Regression Analysis of the Exhaust Temperature for the Two-Stroke-Cycle Diesel Engine

1.1 The heat density of main diesel engine

At present there are many marine ships and boats that widely use the two-stroke-cycle diesel engines as they have the advantages over the four-cycle diesel engines **(Osbourne,1944)**.

However, the question of heat density and functional analysis of the exhaust temperature for these engines have not been investigated enough **(Whalley,1992)** and **(Reader and Hooper, 1983)**.

Some attempts in this question were made by the authors **(Avallone and Baumeister, 1987)** in view of functional analysis of the exhaust temperature in connection with relative load for the two-stroke-cycle diesel engine. These conclusions indicate that with increasing the relative load (or engine speed), the exhaust temperature accordingly increases ,but this regression analysis has the nonlinear relationship.

However, the author of this paper does not agree with such a conclusion and the character of the above-named distribution seeks to investigate in this paper the general questions which are joined with heat density of the two-stroke-cycle diesel engine and exhaust temperature in connection with some of the parameters of a running ship in the tropics such as the seawater temperature, duration in-service of ship, wind speed an direction of wind, ship's speed; and other parameters .

The author thinks that such functional analysis can discover more widely and accurately the complex problems of heat density and the exhaust temperature, which is more important for the diesel engine because the latter works in difficult conditions such as the running ship in the tropics.

The ship is the complex energetic arrangement consisting of the main diesel engine and some auxiliary mechanisms. It is known that the general index of heat density of a diesel engine is the exhaust temperature. However, the conditions of a working engine in the tropics are very difficult because the exhaust temperature rises considerably with the increase of duration in-service of a running ship.

These conclusions are confirmed by statistical results which are shown in Figure 1. Analysis of Figure 1 shows that the exhaust temperature has an irregular character for the period of a running ship in the tropics.

As indicated in Figure 1 ,there is a general pattern in the increase of the exhaust temperature of a diesel engine from the starting point of the operation ship and engine.

1

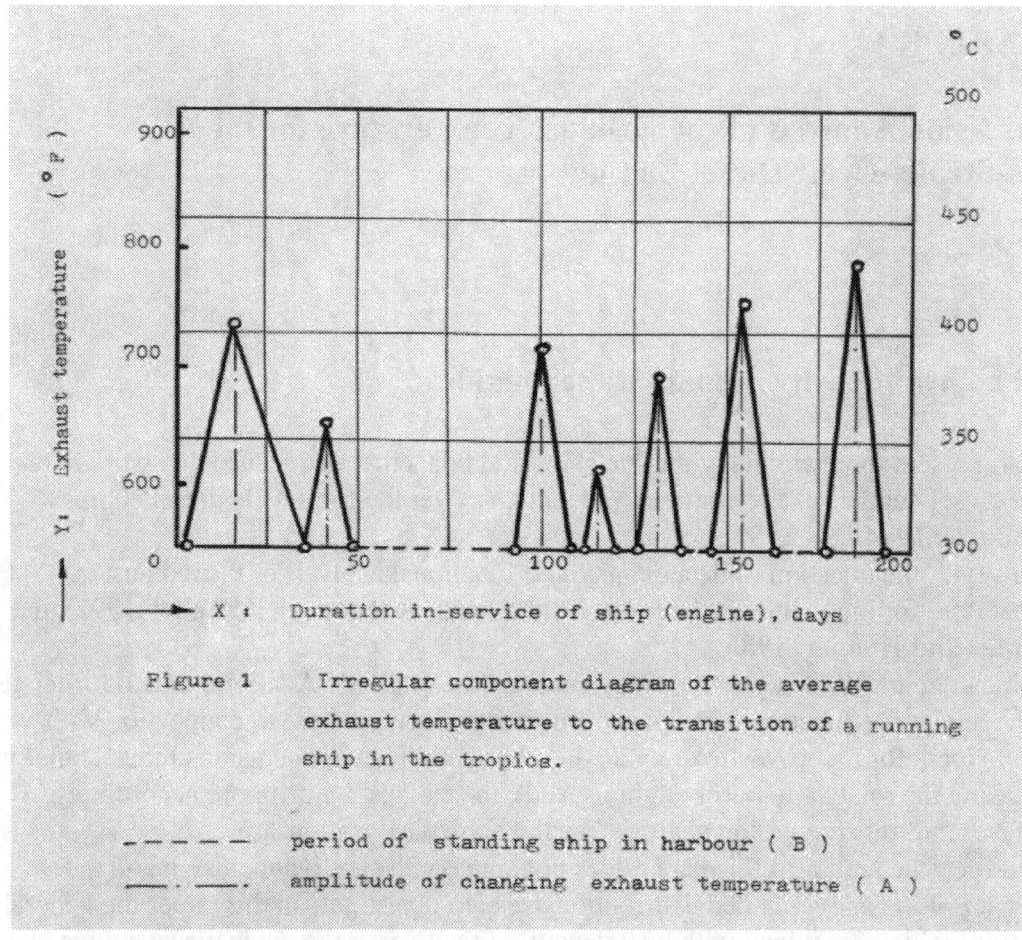

Figure 1   Irregular component diagram of the average
           exhaust temperature to the transition of a running
           ship in the tropics.

- - - - -   period of standing ship in harbour (B)
——— . ———.   amplitude of changing exhaust temperature (A)

The author admits also that Figure 1 in a generalized view is a typical graph for the different diesel engines where the value marked (A) can take the different values and characterize the amplitude of deviation for each phase of this irregular changing of the exhaust temperature for the period of mooring of the ship in harbor or on the open roadstead can characterize the other value which is marked (B) and has the different values for each phase of this distribution for all periods of operation of a ship in the tropics.

As shown in Figure 1 , the function of the exhaust temperature (Y) is the time series model and submits to its laws and rules. These conclusions are confirmed by the data which is shown in Figure 2 in view of the scatter-plot diagram of the exhaust temperature and some average forecasting models for it.

Considering a graph at a given exhaust temperature in Figure 2, we see that the six-month actual average temperature submits to the nonlinear regression model and has the three- parameter curve having a minimum but without an inflection; i.e characterized in a generalized view as quadratic polynomial

$$Y = \alpha + \beta X + \delta X^2 \qquad (1) \qquad \text{where } \delta > 0$$

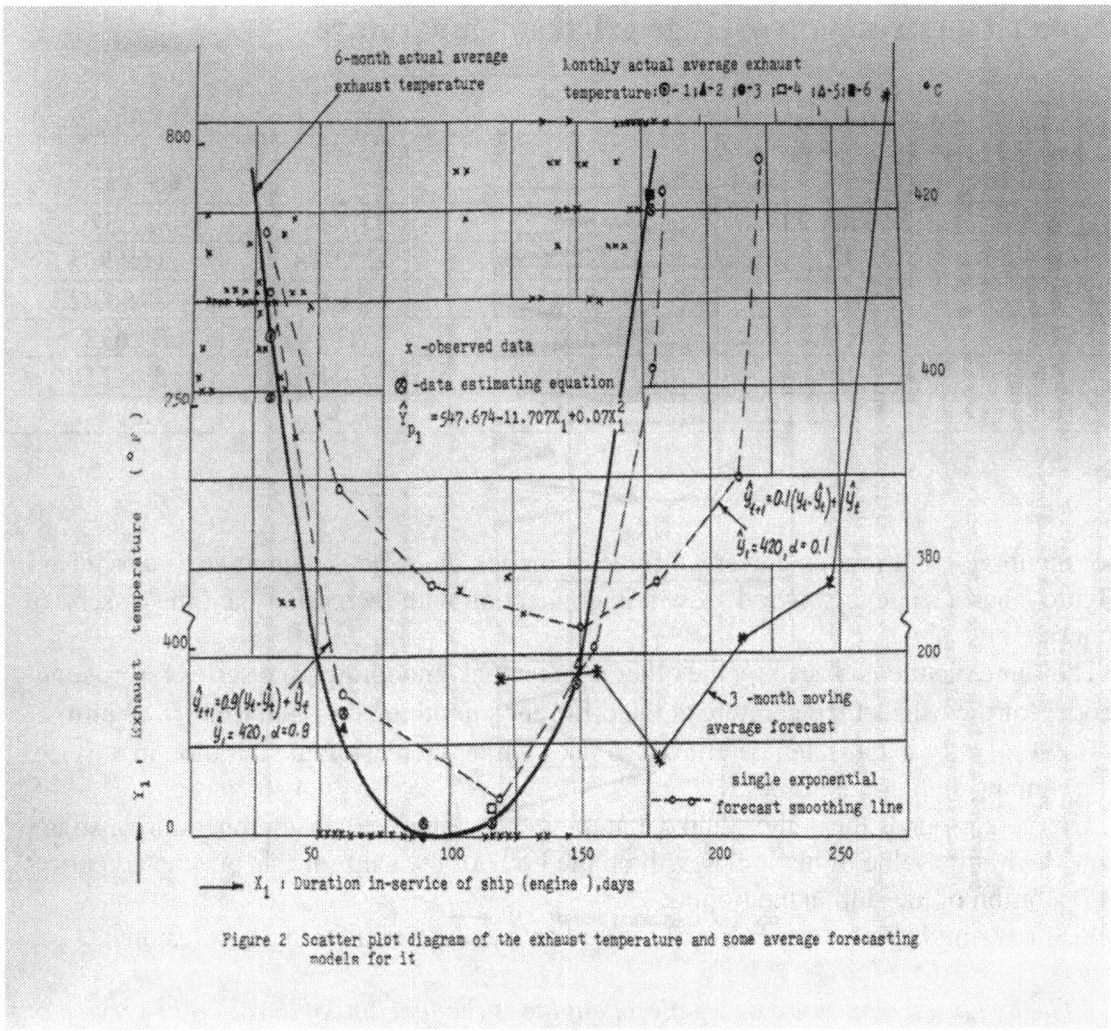

Figure 2 Scatter plot diagram of the exhaust temperature and some average forecasting models for it

Therefore ,it is obvious that the relationship between time in-service of ship (engine) and exhaust temperature is within the curvilinear graph and conforms to the data shown in Figure 2 ;this graph has the estimating equation

$$\hat{Y}_{p,1} = 547.674 - 11.707\, X_1 + 0.07\, X_1^2 \qquad (\,2\,)$$

Table 1 shows the actual and estimated data for the six-month average exhaust temperature conforming to Figure 2.

Table 1 The six-month average exhaust temperature

X_i	Y_i	$Y_{p,1}$	$(Y_i - Y_{p,1})$	$(Y_i - Y_{p,1})^2$
30	408.70	386.846	21.854	477.577
60	142.43	155.428	-12.998	168.955
90	0	41.838	- 41.838	1750.423
120	53.47	46.076	7.394	54.673
150	192.60	168.142	24.458	598.210
180	425.20	424.154	1.046	1.094

For the three-month moving average forecast model, the exhaust temperature as shown in Figure 2 has a tendency toward growth in connection with increasing the time in-service of a running ship.

The same picture is placed by the other exponential smoothing time series forecasting model for the exhaust temperature at the different smoothing constant ($\alpha = 0.10$ and $\alpha = 0.90$); i.e the exhaust temperature also rises while increasing the duration in-service of a running ship in the tropics.

On these grounds the author thinks that the increasing of exhaust temperature also is joined with the value of marine growth on the body of the ship for this protracted period of operation of the ship in the tropics.

So, analyzing Figure 1 and Figure 2, the author makes the following conclusions:

1. *The exhaust temperature of the diesel engine is the function of duration in-service of a running ship in the tropics;*

2. *Regression analysis of the exhaust temperature has the nonlinear character;*

3. *The functional model of the exhaust temperature submits to the time series forecasting models with its laws and rules;*

4. *The character of the average exhaust temperature for all periods of a running ship in the tropics has an irregular component diagram with the multiple of phase and branches;*

5. *In each branch and phase of this diagram ,the exhaust temperature graph has placed some definite value in view of amplitude (A) and the value of period (B) of a mooring ship in a harbor or the open roadstead for this transition;*

6. *The nonlinear character of the changing of the average exhaust temperature versus duration in-service of a running ship (diesel engine) in the tropics presents a generalized view of the typical graph for many marine diesel engines and ships;*

7. *Tendency and natural laws of changing the exhaust temperature is characteristic also so that the value of it considerably rises when increasing the period of operation of a ship (engine) in the tropics And this rising of the exhaust temperature is provoked by the marine growth on the body of a ship which also considerably increases with the duration in-service of a ship in the tropics;*

8. *The average values of the exhaust temperature for the diesel engine are characterized and described in a generalized view as the quadratic polynomial equation view*

$$Y = \alpha + \beta X + \delta X^2 \qquad where \ \delta > 0$$

And this equation will be correct for many marine cargo and passenger ships running in the tropics;

9. *So, the exhaust temperature in the function of the duration in-service of a running ship submits to the time series forecasting model and for this objective possibility to use the moving average forecast models and single exponential forecasts smoothing line for evaluation of exhaust temperature as a forecasting process for running ships(diesel engine) in the tropics;*

10. *Analyzing the above-named conclusions ,the author thinks that the exhaust temperature from the diesel engine (Y_i) is the function of duration (X_i) in-service of a ship(engine) and this functional model submits to the regression analysis view of* $Y_i = \varphi (X_i)$ *and can be described as the nonlinear character of this relationship.*

1.2 Thermodynamic aspects of operation engine in tropics

These problems were discussed by the authors **(Avallone and Baumenstein,1987).** However, the author of this paper , on the basis of statistical data (*observed data I=183) for the cargo ship deidveit =10,984 ton and the two-stroke-cycle diesel engine=8,750 bhp (model " Burmister-Wein") shows that the exhaust temperature has a linear regression function.*

So, this functional analysis has the view $\mathbf{Y_2 = \varphi_1(X_2)}$**.** Figure 3 shows a scatter diagram of the engine speed versus exhaust temperature.

In view of the fact that the coefficient of correlation shown in Figure 3 between the engine speed and exhaust temperature approaches the value of one (r=0.997) ,it may be added to this question that the exhaust temperature submits to the linear regression analysis; i.e the rising of the engine speed directly or proportionally increases the exhaust temperature from the diesel engine. These relationships can be described in view of regression equation:

$$\hat{Y}_{p,2} = 0.34 + 3.92 X_2 \qquad (3)$$

The data of these conclusions are shown in Figure 3 where the good correlation is indicated between the exhaust temperature and engine speed.

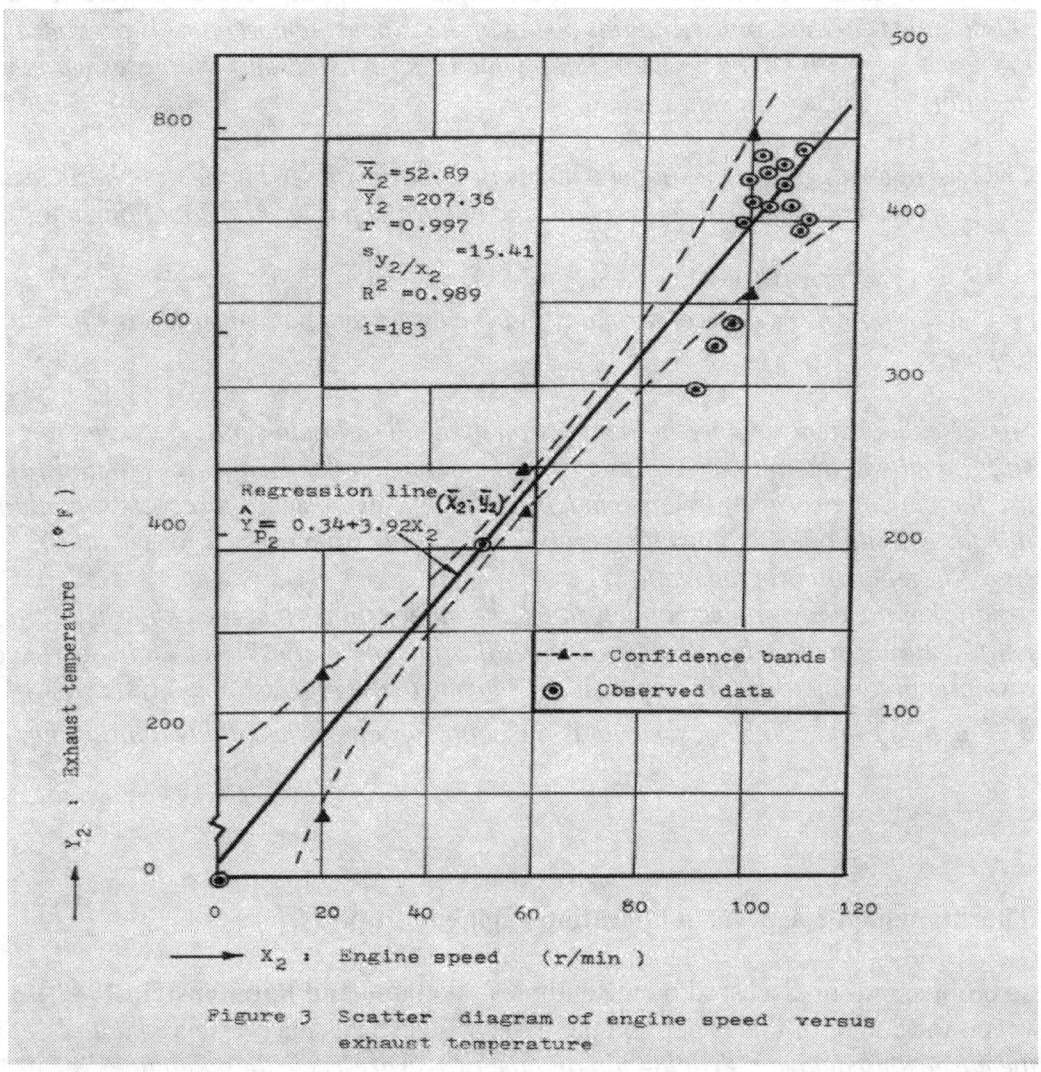

Figure 3 Scatter diagram of engine speed versus exhaust temperature

In view of the fact that the coefficient of determination R^2 indicates that this value is equal to 98.90 percent of the variability in the exhaust temperature (Y_2) so that it may be concluded that a very strong linear relationship has been identified in this regression analysis because all observed data falls perfectly on the fitted regression line

$$\hat{Y}_{p,2} = 0.34 + 3.92X.$$

And the value with a 90 percent confidence interval on $\mu_{y,2/x,2}$ as shown in Figure 3 and given for a set of X_2 value in Table 2 indicates the fact that the value X_2 " moves away " from the value $\bar{X}_2^* = 58.00$;i.e the length of the confidence intervals increases with the increasing of variable value $\hat{Y}_{p,2}$.

Table 2   Ninety percent confidence intervals(*) for $\mu_{y,2/x,2}$ for X_2

X_2	$\hat{Y}_{p,2}$	Lower limit(*)	UpperLimit(*)	Length (*)
20	78.74	38.84	116.64	79.80
40	157.14	117.24	197.04	79.80
—* X_2=58.00	227.70	223.70	231.70	8.00
80	313.30	273.40	353.20	79.80
100	392.30	352.32	432.29	79.97
115	451.14	411.20	491.04	79.84

Figure 4 shows the residual plot for the engine speed using the data shown in Figure 3.

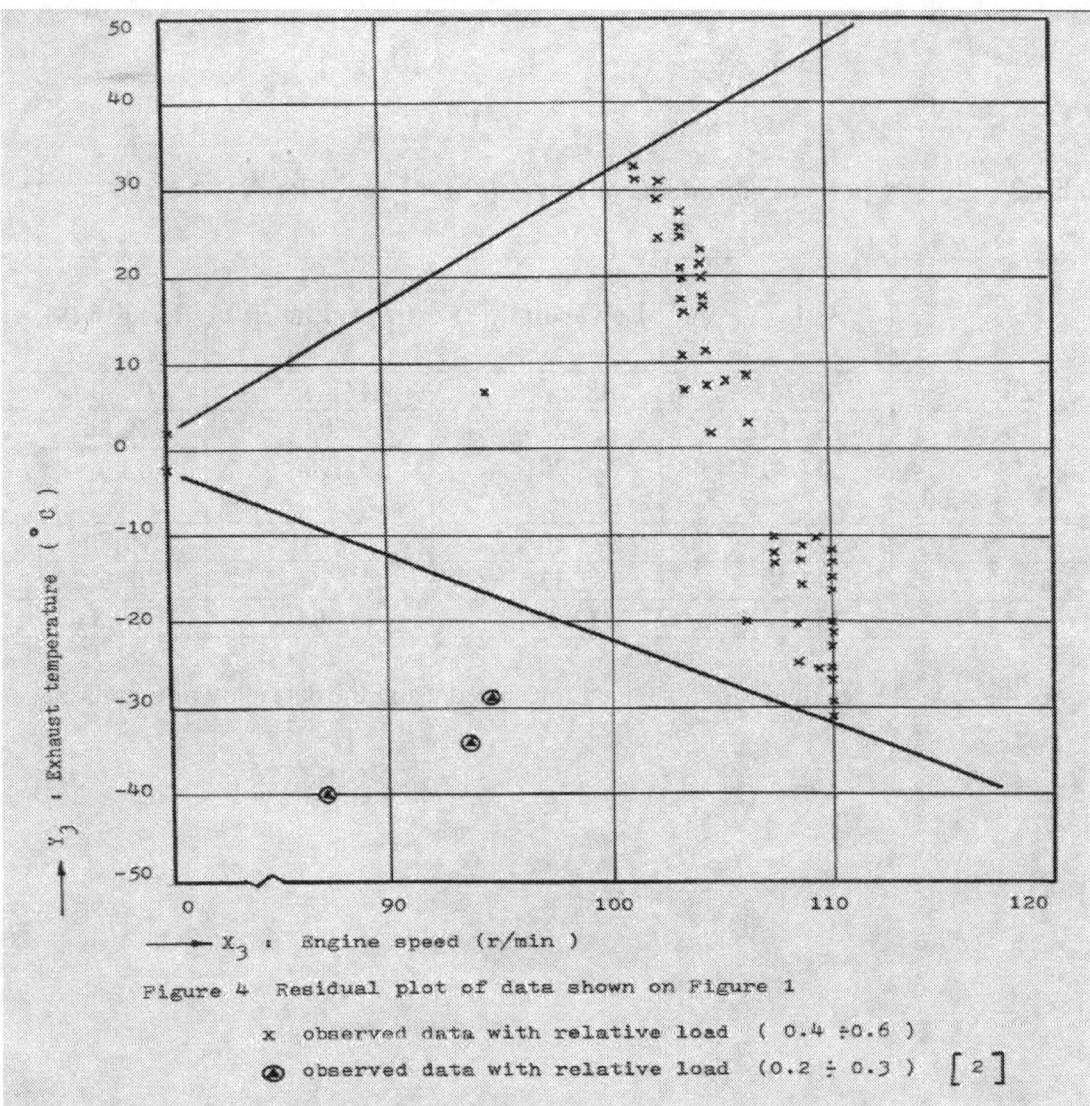

Figure 4   Residual plot of data shown on Figure 1

x   observed data with relative load (0.4 ÷0.6)

⊗   observed data with relative load (0.2 ÷ 0.3)   [2]

The residual encircled in Figure 4 corresponds to the observation $Y_2=300°C$ when $X_2=87$ r/min; $Y_2=345°C$ when $X_2=95$ r/min and $Y_2=335°C$ when $X_2=94$ r/min , Which are shown in Figure 3 indicates the fact that the data of these exhaust temperatures fall to the other relative load of the diesel engine.

The value of it is equal **0.20 ÷0.30** ,distinguishing from the data shown in Figure 3 with the relative load **0.40÷ 0.60** and moves off away considerably from the regression line $\hat{Y}_{p,2} = 0.34 + 3.92\ X_2$. So, the analysis of Figure 3 and Figure 4 allows the following conclusions:

1. *The thermodynamic relationship has the place between the engine speed and exhaust temperature so that this functional connection submits to the regression linear analysis;*

2. *The exhaust temperature and heat density rises accordingly with the increasing of relative load (engine speed);*

3. *Between engine speed (X_2)and the exhaust temperature (Y_2) for the two-stroke-cycle diesel engine there is the linear regression dependence of view*

$$\hat{Y}_{p,2} = 0.34 + 3.92\, X_2$$ *so that this regression line has to place 90 percent confidence interval bands for* $\mu_{y,2/x,2}$ *for* X_2 *values.*

1.3 Influence of the different external factors on the exhaust temperature of engine

a. *The heat density of a diesel engine as a complex function of multiple variables.*

The author thinks that the seawater temperature functionally is joined with the heat density (H_g) of a working diesel engine, and this dependency is the complex function of view $H_g = \psi(Q)$ where Q equals the quantity of heat absorbed by the diesel engine from burning gases in cylinders.

As was indicated by the author (**Osbourne,1944**),the quantity of heat absorbed by the diesel can be removed from it **30 to 35** percent by the cooling system with the use of seawater.

So, the above-named recommendations mark the fact that is the function connection between heat absorbed by the diesel engine and seawater temperature (t_w) in view $Q = \varphi(t_w)$. But the quantity of heat absorbed by the diesel engine which is also functional depends on the exhaust temperature (T_g),i.e this is correct for the functional of view $Q = \varphi_1(T_g)$ where the exhaust temperature is also the function of relative load (engine speed),i.e $T_g = \varphi_2(n)$.

Analysis of the dependency of the exhaust temperature in connection with the duration in-service (N) of a running ship in the tropics ,as was shown in Figure 2 ,also admits that the exhaust temperature has the function of view $T_g = \varphi_3(N)$,i.e the exhaust temperature is the complex function of view $T_g = \varphi_4(n,N)$ or $T_g = \varphi_5\{\varphi_4[\varphi_2(n);\varphi_3(N)]\}$.

Therefore, the complex functional model for quantity of heat absorbed by the diesel engine can be expressed in view of:

$$Q = \varphi_6 \|\varphi_1\{\varphi_5(\varphi_4[\varphi_2(n);\varphi_3(N)])\};\varphi(t_w)\| \qquad (4)$$

And for the heat density of the diesel engine this dependency has the following view:

$$H_g = \psi[\varphi_6 \|\varphi_1\{\varphi_5(\varphi_4[\varphi_2(n);\varphi_3(N)])\};\varphi(t_w)\|] \qquad (5)$$

From the functional equations (4) and (5) we see that the heat density and the exhaust temperature of a diesel engine include such variable as:

n = engine speed (relative load);
t_w = seawater temperature;

9

N= duration in-service of a running ship(engine)in the tropics.

b. *Correlation of the exhaust temperature from a diesel engine and seawater temperature.*

As was indicated above , the exhaust temperature (T_g) is the function of relative load (engine speed) ,and this dependency expresses such an equation as $T_g=\varphi_2(n)$. Such a dependency is confirmed by the data which is shown in Figure 3. However, the exhaust temperature is also the function of seawater temperature, i.e this dependency has view of $T_g=\psi_1(t_w)$ **(6).**

   This functional model (6) is confirmed by the data shown in Figure 5 where it is introduced in the scatter-plot diagram of seawater temperature versus exhaust temperature of a diesel engine.

   Analysis of Figure 5 shows that exhaust temperature has the linear regression character of distribution according to the seawater temperature. Therefore, it can be marked more accurately that the exhaust temperature is the complex functional model of multiple variables and has the following view:

$$T_g=\psi_2 \mid\mid\varphi_5\{\varphi_4[\varphi_2(n);\varphi_3(N)]\};\psi_1(t_w) \mid\mid \qquad (7)$$

The data shown in Figure 5 also indicates that these relationships have a linear regression dependency and are characterized with a negative slope. From this functional model (7) and Figure 5 it may be concluded that with a rising seawater temperature (t_w) the exhaust temperature from the diesel engine accordingly decreases.

   These conclusions are confirmed by the data in Figure 5 for correlation of the exhaust temperature and seawater temperature and is disclosed in view of regression line:

$$\hat{Y}_{p,4} = 1106.436 - 35.387\ X_4 \qquad (8)$$

with estimated data including the standing of the ship (observed data *I=183)* and for the regression line of view   1

$$Y_{p,4} = 440.552 - 1.228\ X_4 \qquad (9)$$

                                                                    1

with estimated data without standing of the ship (observed data **I = 92**) and some characteristics which are shown in Figure 5.

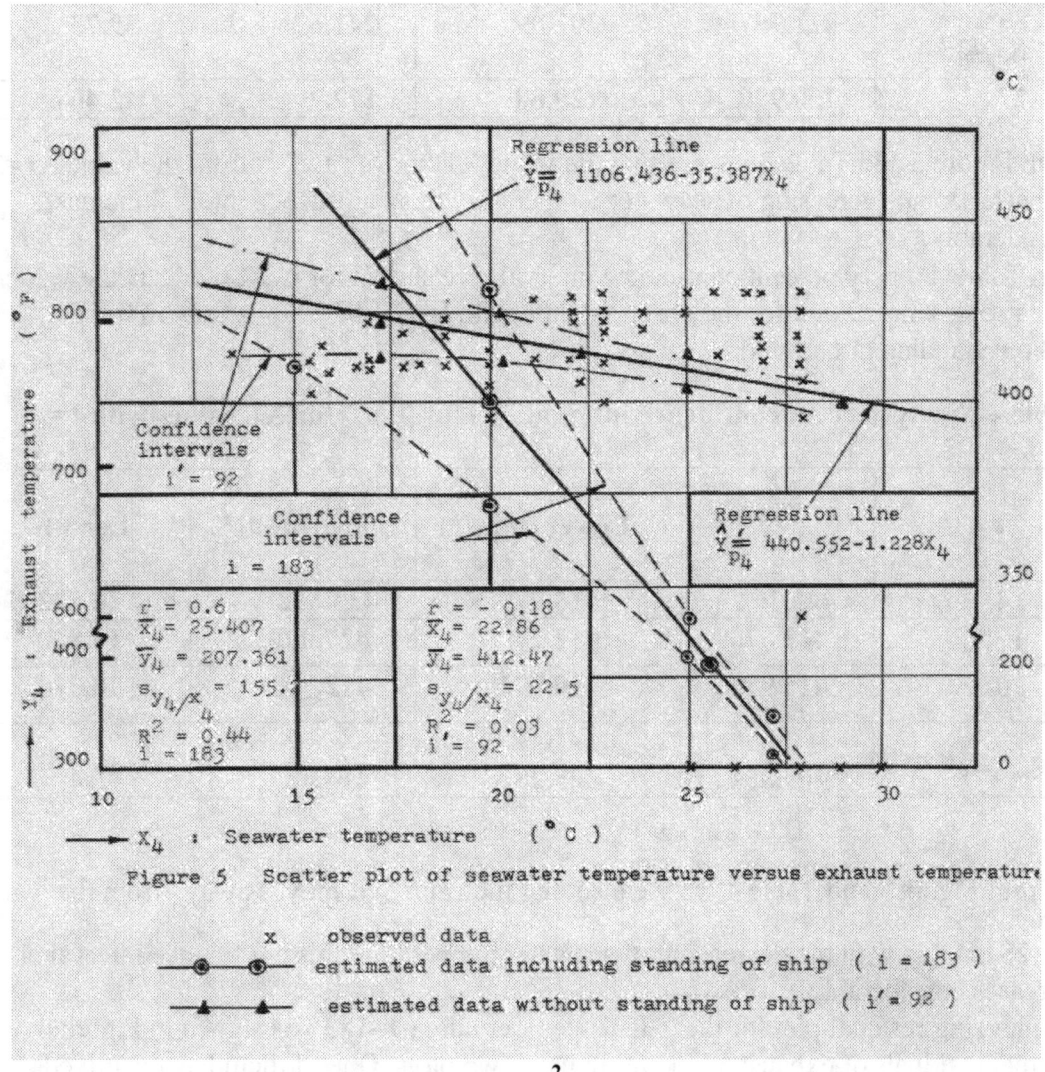

Figure 5   Scatter plot of seawater temperature versus exhaust temperature

At that time the coefficient of determination R^2 for the observed data with **I=183** is equal to 0.44 ,i.e **R=0.44** ,and we see only 44.00 percent of the variability in **Y4**.

Therefore ,there is a middle strength of the linear relationship between the seawater temperature and exhaust temperature. The worst picture we can see is in Figure 5 with the observed data **I=92,** where the coefficient of determination is equal to 0.03,i.e R^2 **=0.03,**so only 3.00 percent of the variability is in **Y4**.

Estimated 90 percent confidence intervals for $\mu_{y,4/x,4}$ for the values of seawater temperature are shown in Table 3 for observed data with **I=183.**

Table 3 Ninety percent confidence intervals(*) for $\mu_{y,4/x,4}$ for X4 values

X_4	$\hat{Y}_{p,4}$	Lower limit(*)	Upper limit(*)	Length
20	398.697	366.36	431.04	64.68

$\overline{}^*$ $X_4 = 25$	221.757	201.89	241.62	39.73
27	150.990	129.80	172.20	42.40

From Figure 5 and Table 3 we see that the value X_4 "moves away "from the value $\overline{X_4}^* = 25$ so that with the decreasing of seawater temperature this confidence interval increases considerably.

In Table 4 ninety percent confidence intervals are shown for $\mu_{y,4/x,4}$ for X_4, the values of seawater temperature for the observed data with $I=92$ (the ship is only running without standing in harbors).

Table 4 Ninety percent confidence intervals (*) for $\mu_{y,4/x,4}$ for X_4 values with $I=92$[1]

X_4	$\hat{Y}_{p,4}^{1}$	Lower limit(*)	Upper limit(*)	Length
14	423.362	413.07	433.652	20.582
17	419.676	412.27	427.090	14.820
20	415.992	410.992	412.002	10.010
$\overline{X_4}^{*1} = 25$	409.852	405.28	414.42	9.14

From Figure 5 and Table 4 we see also that the value X_4 "moves away" from the value $\overline{X_4}^{*1} = 25$ so that with the decreasing of seawater temperature this confidence interval also increases considerably.

Analyzing residual plot for the exhaust temperature ($I=183$) as shown in Figure 6 ,we see that residuals of exhaust temperature have two plots. One of them has the mixed residuals as negative at the beginning of the experimental data; at the end of the experiment, the residuals are positive.

The other plot has only the negative residuals as these observations certainly lead to the condition that the ship for a long time was standing in the harbor and on the roadstead (about of three months) from the total duration of operation of the ship (engine) in the tropics. On the basis of the above-stated facts and the data of Figure 5 and Figure 6, the author makes the conclusion that between seawater temperature and the exhaust temperature there is the correlation of the linear connection with the negative slope so that the value of slope of this fitted regression line $\hat{Y}_{p,4} = 1106.436 - 35.387\, X_4$.

It characterizes itself as the observed data with $I=183$ of the mixed view (the ship is in a regime of transition and standing) considerably larger than the value of slope regression line $\hat{Y}_{p,4}^{1} = 440.552 - 1.228\, X_4$, characterizing itself as the observed data with $I=92$ (the ship is in regime only of transition).

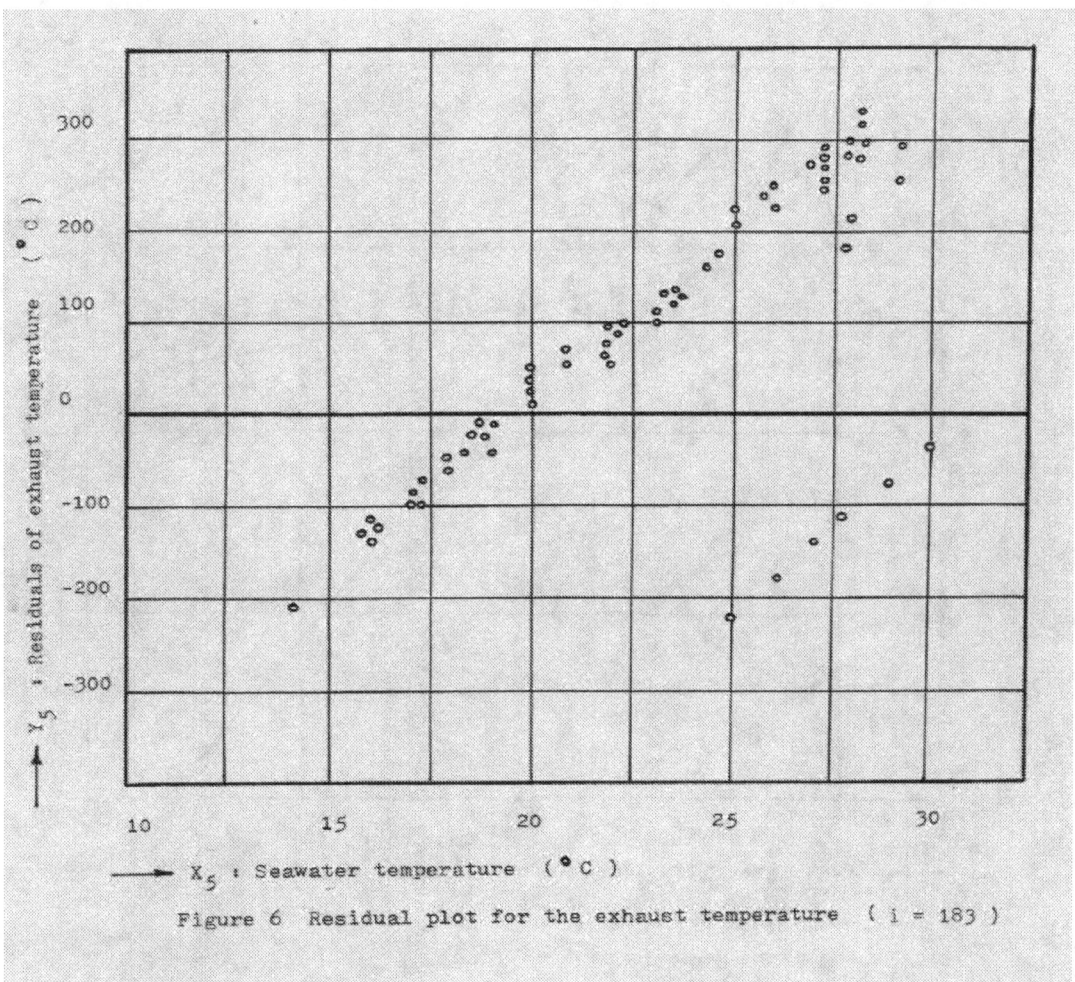

Figure 6 Residual plot for the exhaust temperature (i = 183)

The author makes the conclusion that the periodical or long-time standing of the ship in the harbor or on the roadstead positively influences the heat density of a working engine, i.e with an increase of seawater temperature accordingly decreases for the exhaust temperature from the engine , and this all can be reached by the fact that diesel engine did not operate for a long time in the tropics ,particularly on the standing processes with observed data **I=183** and the coefficient of determination is equal to $\mathbf{R^2} =\mathbf{0.44.}$

The different pictures are placed with observed data $\mathbf{I_1=92}$ (the ship only running), and we see that the coefficient of determination $\mathbf{R^2}$ is equal to $\mathbf{R^2} =\mathbf{0.03,}$ i.e the author thinks this has placed a small correlation between seawater temperature and the exhaust temperature as this is shown in Figure 5 with insignificance negative slope (coefficient b_1 is equal to $\mathbf{b_1 = - 1.228).}$

The conclusion is made by the author about changing engine speed from seawater temperature as shown in Figure 7.

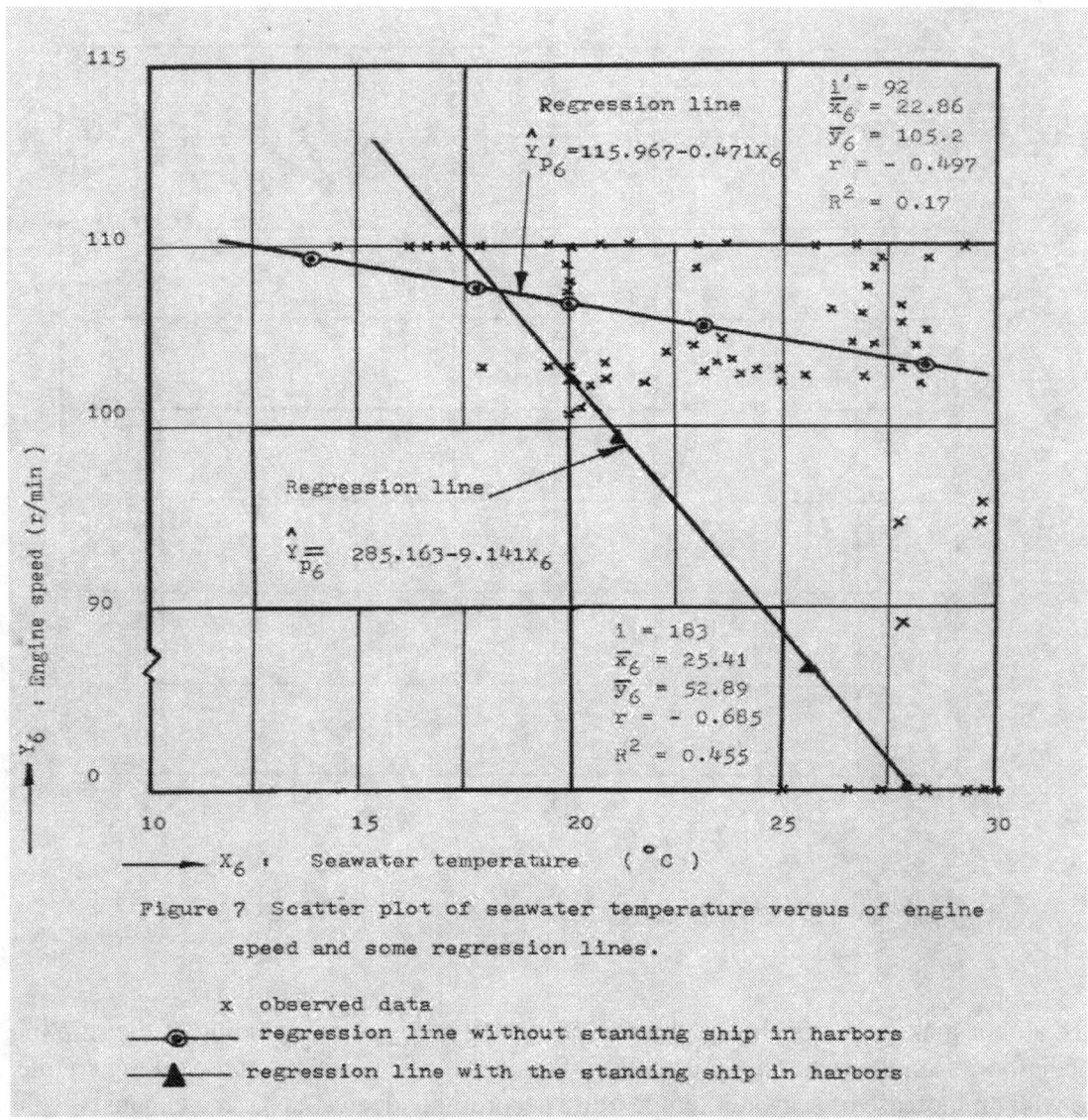

Figure 7 Scatter plot of seawater temperature versus of engine
speed and some regression lines.

x   observed data

———⊚——— regression line without standing ship in harbors

———▲——— regression line with the standing ship in harbors

This functional analysis shows that with increasing seawater temperature the engine speed considerably decreases, and therefore the author's conclusion is confirmed by the data shown in Figure 5 and Figure 7 because the engine speed and the exhaust temperature are joined functionally.

c. *Influence by the other operational parameters of a running ship (wind speed and its direction, ship's speed, etc.) on the character of changing exhaust temperature from diesel engine.*

As was indicated in the above-stated conclusion, the exhaust temperature from the diesel engine is a function depending on multiple factors such as the relative load (engine speed), duration in-service of a running ship, or marine growth of body ship, and also the seawater temperature. But examining more attentively the other operational parameters of a running ship, the author makes the conclusion, on the basis data in Figure 8 ,where it

14

shows scatter plots of wind speed versus the exhaust temperature, illustrating that the functional relationship is absent between the above named parameters as the regression line is perfectly horizontal ,as illustrated in Figure 8.

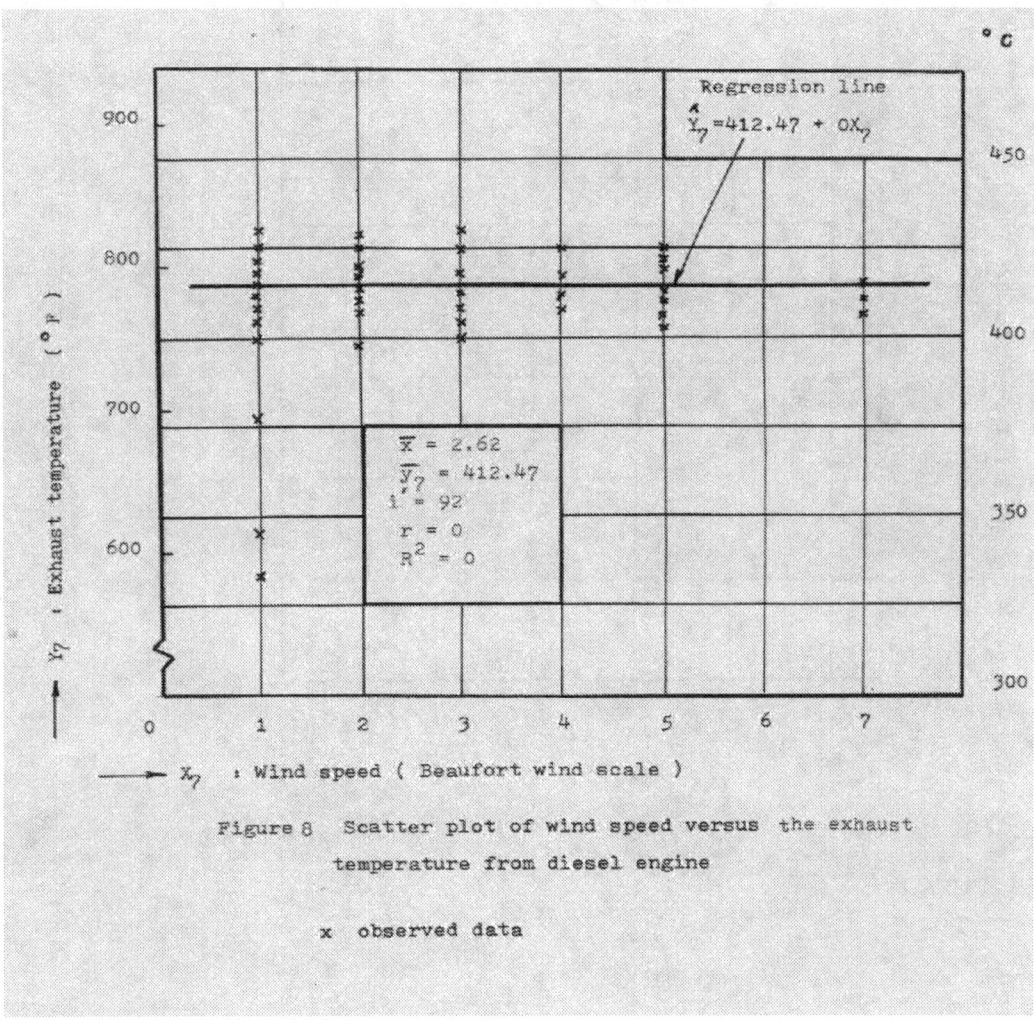

Figure 8   Scatter plot of wind speed versus the exhaust
temperature from diesel engine

x   observed data

Therefore, the value of the coefficient is equal to **$b_1=0$** and the regression line is equal

$$\hat{Y}_{p,7}=412.47 +0X_7 \quad (\,10\,).$$

So , the author thinks that exhaust temperature does not depend on the wind speed and its direction in the process of a running ship in the tropics, i.e the correlation between these values is absent.

These conclusions are confirmed by the data illustrated in Figure 9 ,where shown also are scatter plots of the direction of wind on the body of a ship versus the exhaust temperature, and this dependence is marked by the regression line in view of

$$\hat{Y}_{p,8}=412.47 + 0X_8 \quad (\,11\,)$$

The functional analysis of ship speed and exhaust temperature can be expressed by the dependency of view $Y_9 =\varphi(\,X_9\,).$

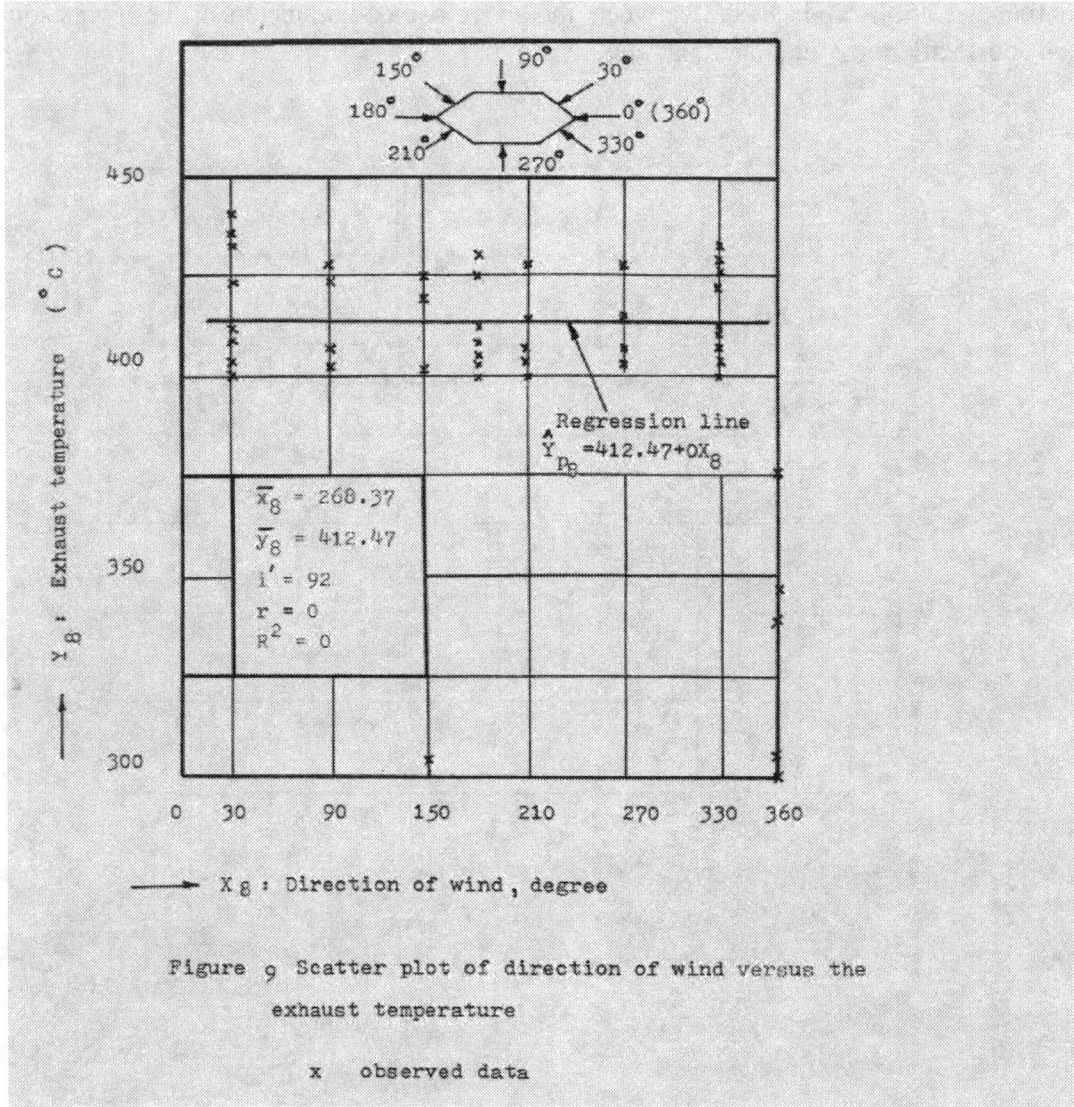

Figure 9 Scatter plot of direction of wind versus the exhaust temperature

x   observed data

As shown in Figure 10, this correlation between the above-stated parameters are marked in view of a scatter plot and the fitted regression line has view $\hat{Y}_{p,9}$ =473.43-4.41X₉ (12) for the exhaust temperature.

In view of the fact that the coefficient **b₁** of this regression line (12) is the negative (**b₁= - 4.41**) , this indicates the fact that the regression line decreases as the value of ship speed **X₉** increases.

Analysis of this function is graphed in Figure 10. The author admits that with the increasing of ship speed, the exhaust temperature accordingly decreases ,and this may be explained by the fact that with increasing ship speed ,the value of marine growth on the body of the ship considerably decreases, i.e accordingly the movement-resistance coefficient also decreases.[7]

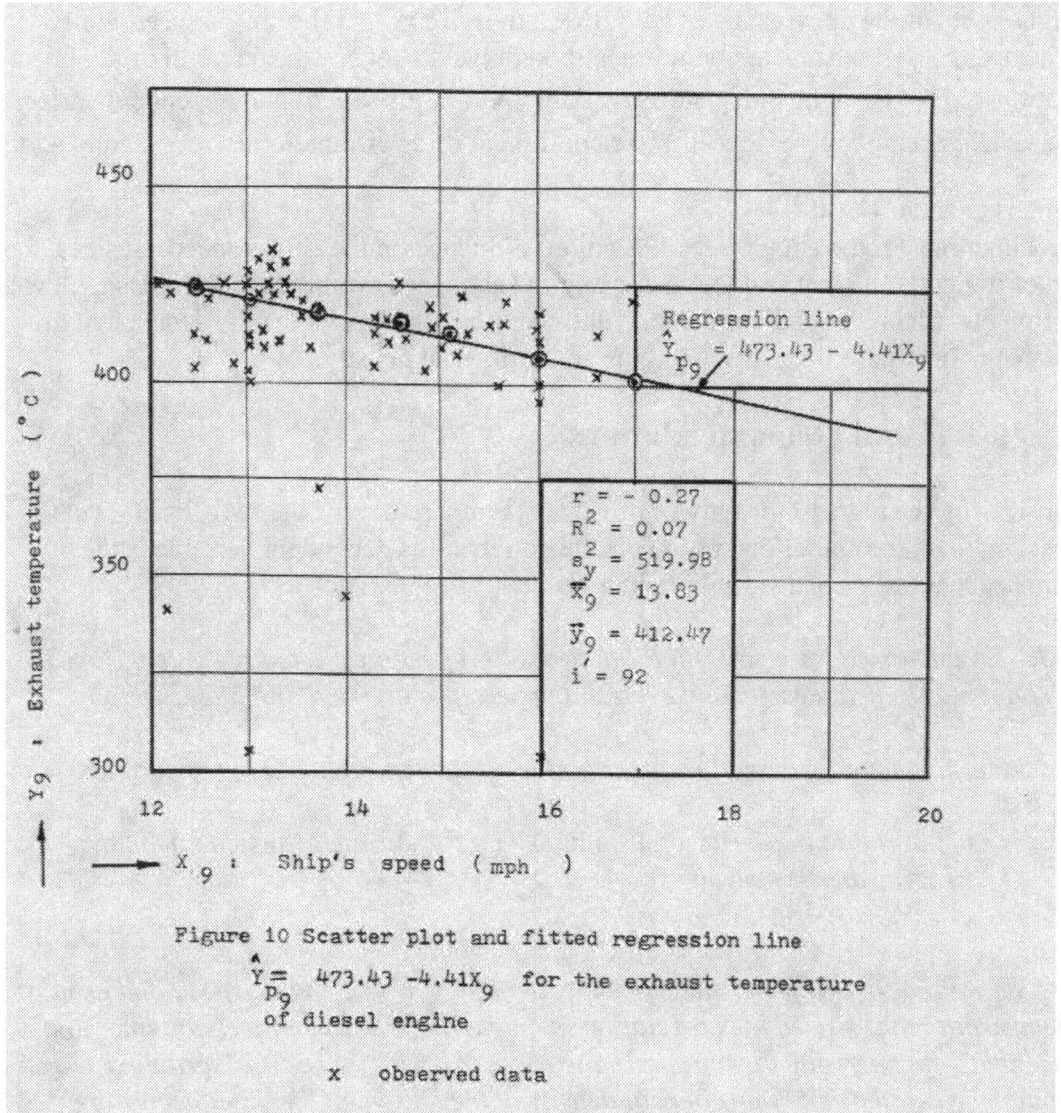

Figure 10 Scatter plot and fitted regression line
$\hat{Y}_{P_9} = 473.43 - 4.41X_9$ for the exhaust temperature
of diesel engine

x   observed data

Therefore, both of these above-named factors improve the work of the diesel engine into account of decreasing heat density and exhaust temperature in the operation processes of both- the ship and engine in the tropics.

This functional connection can be expressed as $X_9 = \varphi_1(X_{10}; X_{11})$ where X_{10} equals the value of marine growth on the body of the ship; X_{11} equals the movement-resistance coefficient and as the above-stated $X_9 = \varphi_2(n)$, where n equals engine speed (or relative load).

So, this functional model has view :

$$X_9 = \varphi_3[\varphi_1(X_{10}; X_{11}); \varphi_2(n)] \quad \text{and then}$$

$$Y_9 = \varphi_4 \|\varphi\{\varphi_3[\varphi_1(X_{10}; X_{11}); \varphi_2(n)]\}\| \quad (13)$$

or $$T_g = \varphi_4 \|\varphi\{\varphi_3[\varphi_1(X_{10}; X_{11}); \varphi_2(n)]\}\| \quad (13\,a)$$

On the basis of the above-stated views, the author thinks that the ship's speed indirectly influences on the exhaust temperature, although we do not see the strength of the linear relationship between the independent variable X_9 (ship's speed) and dependent variable Y_9 (exhaust temperature) because the coefficient of determination is very small and its value is equal to $R^2 = 0.07$.

Functionally, the engine speed has more influence on the ship's speed. Besides, the author indicates the fact that the increasing of ship speed promotes the decreasing level of marine growth on the body of the ship and therefore also promotes the decreasing of exhaust temperature from diesel engine and of its heat density.

Conclusion and recommendations

Considering the regression analysis of exhaust temperature for the two-stroke-cycle diesel engine , the author makes the conclusion that this dependent variable is the complex function of many multiple independent variables such as:

1.   *Engine speed (or relative load)appears as a general thermodynamic parameter changing the heat density and exhaust temperature of a working diesel engine;*

2.*External factors inderectly appear on the character of changing temperature such as :*
   a)   *duration in-service of a running ship (or marine growth on its body);*
   b)   *seawater temperature;*
   c)   *ship's speed.*

   3.*The author in this paper indicates the fact that between engine speed and exhaust temperature has placed the correlation in the view of a linear regression line. And we see that with the rising of ship speed, the exhaust temperature considerably decreases and improves the conditions of heat density of the diesel engine, i.e it accordingly increases its service life because, in this period of running the ship, the level of marine growth on its body is insignificant ;*

4. *And besides we see also that the changing of the average exhaust temperature connected to the in-service duration of a running ship submits to the quadratic polynomial equation of view $Y = \alpha + \beta X + \delta X^2$ ($\delta > 0$) ,and the exhaust temperature for this operation period is characterized as increasing,i.e with the rising of the duration in-service of a running ship ,the exhaust temperature of the diesel engine accordingly increases.*

5. *The author also admits that between wind speed and its direction and also exhaust temperature of diesel engine the correlation is absent;*

6. In this paper the author also indicates the fact that engine speed and exhaust temperature decrease with increasing of seawater temperature because the above-named factors are joined functionally with the seawater temperature.

References

1.Osbourne,*Modern Marine Engineer's Manual,*(New York: Cornell Maritime Press,1944),2:15-15,15-243.

2.P.B.Whalley,*Basic Engineering Thermodynamics,* (New York: Oxford University Press,1992): 203-205.

3.G.T.Reader and C.Hooper, *Stirling Engines, (London: E.&F.N.Spon.,1983):88-89.*

4.Eugene A.Avallone and Theodore Baumeister III ,*Mark's Standard Handbook for Mechanical Engineers,(* New York : McGraw-Hill Book Company,1987): 9-112.

Bibliography

Bowerman,Bruce L. and O'Connell,Richard T., *Time Series and Forecasting.*North Scituate, Massachusetts: Duxbury Press,1979

Croxton,Frederick E.and Cowden,Dudley J., *Applied General Statistics.* New York: Prentice-Hall,Inc.1939

Plaffenberg, Roger C. and Patterson,James H.,*Statistical Methods.* Homewood, Illinois: Richard D.Irwin,Inc.,1977

Ratkowsky,David., *Handbook of Nonlinear Regression Models.*New York : Marcel Dekker,Inc. 1990

CHAPTER 2

Multiple Regression Analysis in Evaluation of Main Diesel Engine Parameters for the Tropical Seawaters

2.1 Brief peculiarities of operation marine engine in the tropics

Statistical methods widely use in the different areas of science and industry, particularly in areas which demand the big investments on realization of experimental and research works.

There are many examples in the mathematical literature use of these methods in biology and life science investigations.

The main purpose of this paper to show the peculiarities of operation the marine diesel in condition of sailing ship in the tropical seawaters and wave sea.

And besides ,on the basis of statistical observation data ,using the multiple regression methods ,to show the functional dependence of engine speed of two-stroke-cycle diesel from some external factors such as : seawater temperature ,wave sea (wind speed and its directions) and operation of ship (or its fouling).

In the process a long period time of operation the main marine diesel engine, particularly in the condition of operation in tropics, has a place some peculiarities of work of this engine:

1. *It is known that the exhaust temperature of diesel engine is the general criterion of its heat density. As rule, this parameter increases with the increasing of operation service of diesel engine in the tropical seawaters. The exhaust temperature value sometimes exceed **400 °C to 430 °C** for two-stroke-cycle marine diesel engine;*

2. *The seawater ,using for inner cooling system of diesel engine in the tropics, does not guarantee the good operational conditions because the temperature of seawater is very high and its value sometimes exceed **25 °C to 32 °C**;*

3. *In period of sailing ship in the tropics the engine speed of diesel decreases, particularly with increasing of operational period. In these conditions, the engine speed periodically changes by the operational people that to support the normal heat density of marine diesel engine;*

*As rule, the high exhaust temperature of diesel engine at this time and necessity its of decreasing with assistance of reducing the engine speed in the operational process is the main index which shows that body of ship has the big fouling. The average engine speed usually drops and its value does not exceed **10 percent** for the ship, sailing in the tropics more than six months .*

4. *The operation of marine diesel also becomes worse in the result of presence of high air humidity what makes the supercharge system of engine not so productivity;*
5. *It is shown by observations that wave sea (wind speed and its directions) make worse the operation of diesel engine in the tropics ,i.e increases its heat density and decreases the service life of engine.*

2.2 The general directions of selection and processing of statistical data

In base of present research work for the two-stroke-cycle marine diesel engine is put the following conditions:

a) *All statistical data is observed in operational conditions of dry- cargo ship" Balashiha" (Black Sea Navigation Company, Odessa ,Ukraine) in 1967,having the main parameters:*
   - Deidveit=10,984 ton;
   - Type of marine diesel engine = two-stroke-cycle;
   - Horsepower of diesel engine =8,750;
   - Average ship speed =14,25 Knots;
   - Electro-capacity =3,700 Kw;
   - Running area (Black Sea-Indian Ocean and other tropical Seas).

b) *Statistical observed data of **I=183 (see Appendix 1)** are treated by the methods of mathematical statistics;*

c) *In base of independent variables ,as main external factors ,acting on the operational process of diesel engine ,are investigated the following parameters:*

X_1'= seawater temperature ,°C;

X_2' = duration-in service of ship (engine),days;

X_3'=wind speed (Beaufort wind scale);

X_4'=direction wind, degree.

d) In base of dependent variable is put the engine speed which is the main index of thermodynamic process and heat density of diesel engine, i.e parameter **(Y)**.

e) On the basis of functional analysis ,having the following characteristics view of :

$$Y_1 = \varphi\, (X_1') ;$$

$$Y_2 = \Psi\, (X_2') ;$$

$$Y_3 = \alpha\, (X_3') ;$$

$$Y_4 = \gamma\, (X_4') ;$$

and $\quad Y = (X_1' ; X_2' ; X_3' ; X_4')$ are made some conclusions and recommendations.

2.3 Statistical results and discussion

a) Influence of seawater temperature on the operational process of diesel engine

Figure 1 shows a dependence of engine speed from seawater temperature in different duration-in service periods of operation diesel engine in the tropics.

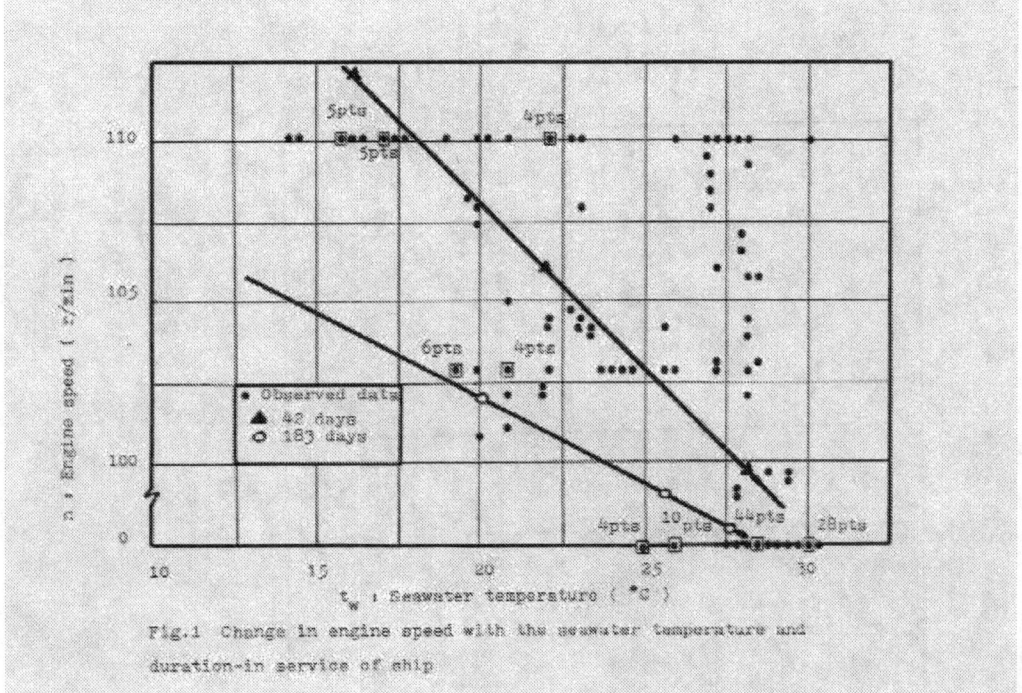

Fig.1 Change in engine speed with the seawater temperature and
duration-in service of ship

It is seen from Figure 1 that in both cases the engine speed decreases with increasing of seawater temperature and duration-in service.

And besides as we see from Figure 1 the value of engine speed is smaller for the case when the duration-in service is larger.

The residual plot of engine speed against seawater temperature for the different duration-in service of diesel engine as shown in Figure 2.

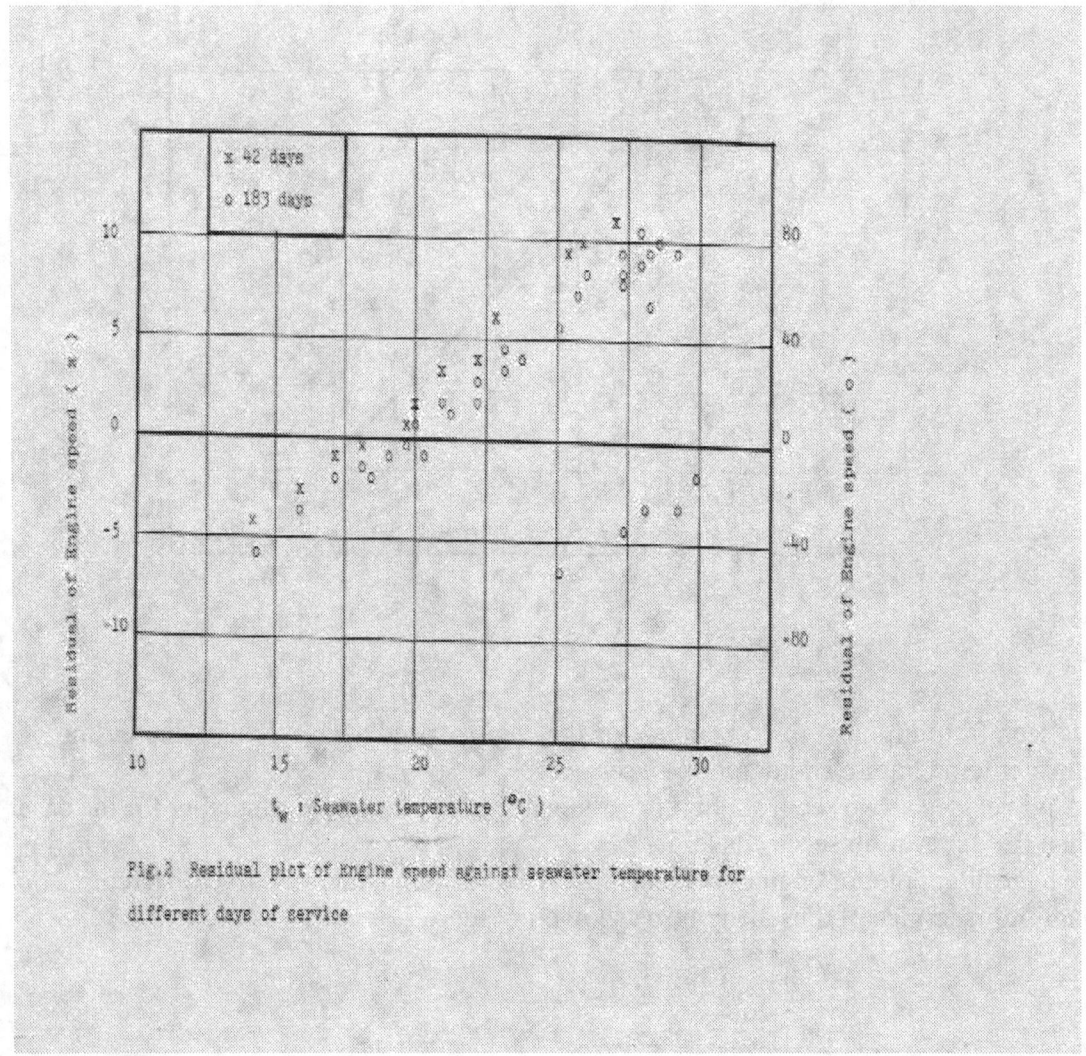

Fig.2 Residual plot of Engine speed against seawater temperature for different days of service

The residual plot from Figure 2 shows that functional dependence view of

$Y_1=\varphi(X_1)$ has linear model. The analysis of variable which is obtained from the different observed data is given in Table 1 – ANOVA table for seawater temperature-engine speed.

Table 1   ANOVA table for seawater temperature-engine speed for the different data

Source of variation	$\hat{Y}_{1,I}=130.92 - 1.16 X_{1,I}$ (I=42)			$\hat{Y}_{1,I}=285.13 - 9.14 X_{1,I}$ (I'=183)		
	Sum of squares (SS)	Degrees of free-dom(df)	Mean square (MS)	Sum of squares (SS)	Degree of free-dom(df)	Mean square (MS)
Regression	1209.16	1	1209.16	231532.3	1	231532.30
Residual	11220.73	40	280.52	273908.1	181	1513.30
Total	12429.89	41		505440.4	182	
Coefficient of correlation	r= - 0.31			r' = -0.68		

It is seen from Table 1 that the coefficient correlation (r= -0.31) is smaller for observed data of **I=42** than for the observed data of **I=183** as the average seawater temperature in this case is insignificance ($\overline{X}_{42}$= **21.88 °C**).

So, we can conclude that with the increasing of seawater temperature, the fitted regression linear model better describes the relationship between the value $X_{1,I}$ and $Y_{1,i}$,when the coefficient of correlation is equal r = -0.68 and in this case the average seawater temperature is equal $\overline{X}_{183}$= **25.41°C** .

In Table 2 and Table 3 are shown the calculations of confidence intervals, tests of hypotheses and prediction intervals for the seawater temperature-engine speed for observed data of **I=42 and I'=183.**

Table 2 Evaluation of linear model $Y_{1,1}=130.92 - 1.16\, X_{1,i}$ (i=42)

Parameter	100(1-α)% confidence interval	Hypothesis H0	t-Test
β_1	$\alpha = 0.05$ $(\hat{\beta}_1 \pm t_{i-2,1-\alpha/2}[S_{y/x}/(S_x(i-1)]^{0.5})$ i= 42; $\hat{\beta}_1$= - 1.16; $\hat{\beta}_0$=130.92; S_x=4.70; $S_{y/x}$=16.75; $t_{40,0.0975}$=2.021; or $-0.03 \leq \beta_1 \leq -2.29$	$\beta_1 = \beta_1^{(0)}$ $\beta_1^{(0)} = 0$	$t^*= (\hat{\beta}_1 - \beta_1^{(0)})\,[S_x(i-1)\ S_{y/x}]^{0.5}$ t^*= -2.076 **Reject H0 at α=0.05** **(two-tailed test)since** $|t^*| \geq t_{i-2,1-\alpha/2}$
β_0	α=0.05 $(\hat{\beta}_0 \pm t_{i-2,1-\alpha/2}\, S_{y/x} \cdot$ $\cdot [1/i + \bar{X}^2/(i-1)S_x^2]^{0.50})$ $\bar{X}$=21.88 **or** $105.77 \leq \beta_0 \leq 156.07$	$\beta_0 = \beta_0^{(0)}$ $\beta_0^{(0)} = 100$	$t^*= (\hat{\beta}_0 - \beta_0^{(0)})/S_{y/x}[1/i + \bar{X}^2/(i-1)S_x^2]^{0.50}$ t^*= 2.485 **Reject H0 at α=0.05** **(two-tailed test) since** $t^* \geq t_{i-2,1-\alpha/2}$
$\mu_{y/x(0)}$ $\mu_{y/19}$	α=0.10 $[\bar{Y}+\hat{\beta}_1(X_0-\bar{X})] \pm t_{i-2,1-\alpha/2} \cdot S_{y/x} \cdot$ $\cdot [1/i + (X_0 - \bar{X})^2/(i-1)S_x^2]^{0.50}$ X_0 =19; $\bar{Y}$=105.64; $t_{40,0.95}$ =1.684; or $103.90 \leq \mu_{y/19} \leq 114.10$	$\mu_{y/x(0)}=$ $\mu_{y/x(0)}^{(0)}$ $\mu_{y/19}=$ 102	$t^* =[\bar{Y} + \hat{\beta}_1(X_0-\bar{X}) - \mu_{y/x(0)}] /$ $/ S_{y/x}[1/i + (X_0-\bar{X})^2/(i-1)S_x^2]^{0.50}$ t^* =3.34 **Reject H0 at α=0.10** **(two-tailed test) since** $t^* \geq t_{i-2,1-\alpha/2}$
Y_{19} Prediction interval at X_0= 19	α=0.10 $[\bar{Y}+\hat{\beta}_1(X_0-\bar{X})] \pm t_{i-2,1-\alpha/2}\, S_{y/x} \cdot$ $\cdot [1+1/i + (X_0 - \bar{X})^2/(i-1)S_x^2]^{0.50}$ $t_{40,0.95}$ =1.684 or $83.36 \leq Y_{19} \leq 140.70$		

Table 3 Evaluation of linear model $Y_{1,i} = 285.13 - 9.14 X_1$ (i=183)

Parameter	100 (1-α) % confidence interval	Hypothesis H_0	t -Test
β_1	α=0.05 $(\hat{\beta}_1 \pm t_{i-2,1-\alpha/2} [S_{y/x}/ S_x (i-1)^{0.50}])$, $\hat{\beta}_1 = -9.14$; $\hat{\beta}_0 = 285.13$ i=183; $S_x = 3.90$; $S_{y/x} = 38.90$; $t_{181,0.975} = 1.973$ or $-7.68 \leq \beta_1 \leq -10.60$	$\beta_1 = \beta_1^{(0)}$ $\beta_1^{(0)} = 0$	$t^* = (\hat{\beta}_1 - \beta_1^{(0)}) S_x(i-1)^{0.50} / S_{y/x}$ $t^* = -12.34$; Reject H_0 at α=0.05 (two-tailed test) since $|t^*| \geq t_{i-2,1-\alpha/2}$
β_0	α=0.05 $(\hat{\beta}_0 \pm t_{i-2,1-\alpha/2} S_{y/x}[1/i + \bar{X}^{-2}/(i-1)S_x^2]^{0.5})$ $\bar{X} = 25.41$ or $247.52 \leq \beta_0 \leq 322.74$	$\beta_0 = \beta_0^{(0)}$ $\beta_0^{(0)} = 240$	$t^* = (\hat{\beta}_0 - \beta_0^{(0)})/ [S_{y/x}(1/i + \bar{X}^{-2} / (i-1) S_x^2]^{0.50}$ $t^* = 2.38$ Reject H_0 at α=0.05 (two-tailed test) since $t^* \geq t_{i-2,1-\alpha/2}$
$\mu_{y/20}$	α=0.10 $\bar{Y} + \hat{\beta}_1(X_0 - \bar{X}) \pm t_{i-2,1-\alpha/2} S_{y/x} [1/i + (X_0 - \bar{X})^2/(i-1)S_x^2]^{0.50}$ $X_0 = 20.00$; $\bar{Y} = 52.89$; $t_{181,0.95} = 1.653$ or $94.62 \leq \mu_{y/20} \leq 110.06$		$t^* = \bar{Y} + \hat{\beta}_1(X_0 - \bar{X}) - \mu_{y/x0}^{(0)} / S_{y/x} [1/i + (X_0 - \bar{X})^2/(i-1)S_x^2]^{0.5}$ $t^* = 2.64$ Reject H_0 at α=0.10 (two-tailed test) since $t^* \geq t_{i-2,1-\alpha/2}$
Y_{25} Prediction Iterval at $X_0 = 20$	α=0.10 $\bar{Y} + \hat{\beta}_1(X - \bar{X}) \pm t_{i-2,1-\alpha/2} S_{y/x} [1 + 1/i + (X_0 - \bar{X})^2/(i-1)S_x^2]$ $t_{181,0.95} = 1.653$; or $37.40 \leq Y_{20} \leq 167.28$		

2.4 Connection of engine speed with wave sea (wind speed and its directions)

The influence of wind speed and its directions on the engine speed of marine diesel is shown in Figure 3.

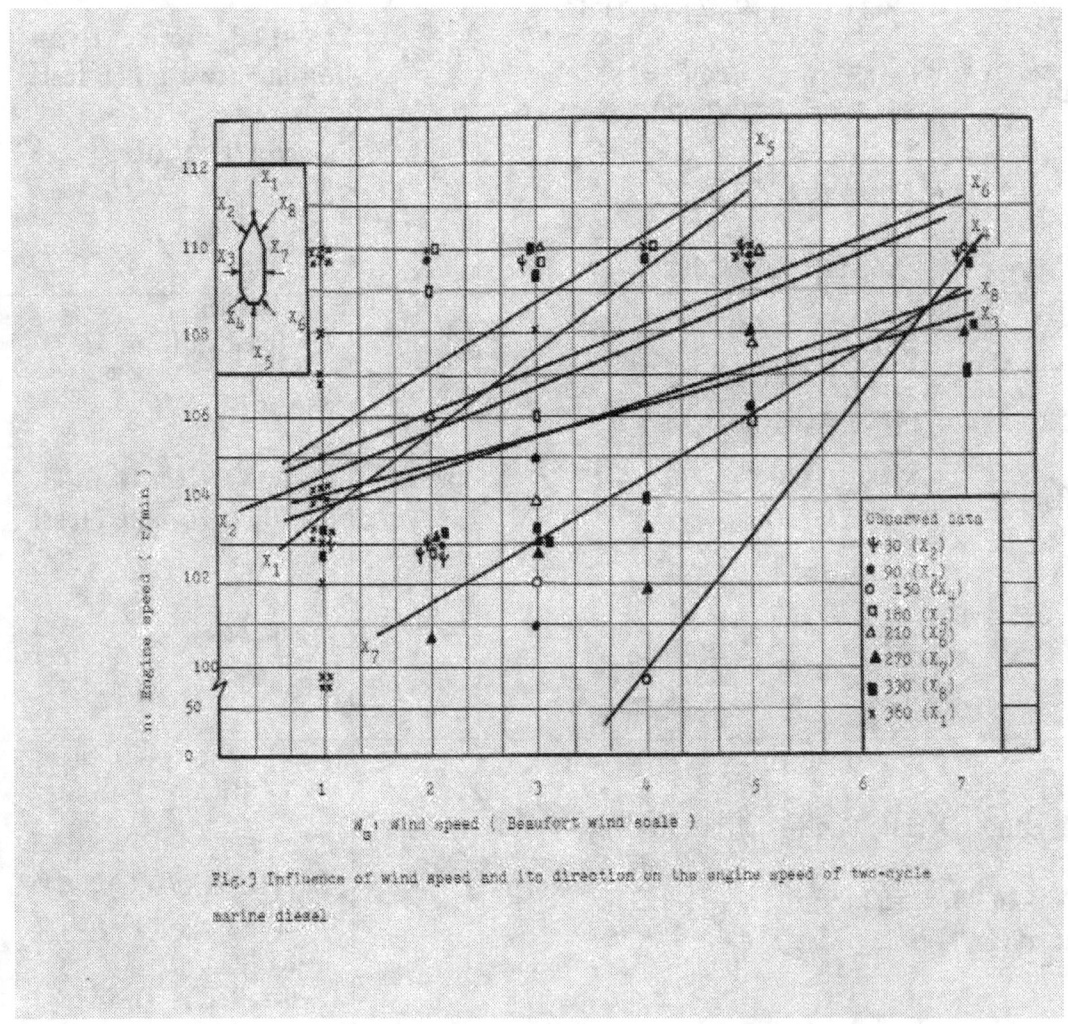

Fig.3 Influence of wind speed and its direction on the engine speed of two-cycle marine diesel

Figure 3 shows that the functional analysis $Y_3 = \alpha(X_3')$ and $Y_4 = \gamma(X_4')$ has linear regression model for any wind directions ,acting on the ship in the tropics.

And besides as shown in Figure 3 the engine speed demands in the operational processes the decreasing for any wind directions. If the engine speed left the constant at this time, its heat density and exhaust temperature will be increased, i.e the conditions for the operation of diesel engine in the tropics and wave sea will be worse.

On the other hand we also see that influence of wind direction on the engine speed is important factor. With this fact should the engine speed to decrease accordingly with acting wind direction on the ship, sailing in the tropics.

Figure 3 also shows that at the wind speed, for instance 4 Beaufort ,the engine speed considerably decreases in any wind directions ,besides the wind direction X_5 when the ship moves with a tail wind.

So, the observed data from Figure 3 show that the wave sea (wind speed and its directions) has the negative influence on the operation of diesel engine particularly in the conditions when the ship sailing in the tropics.

As indicated above, in practical situations should to decrease the engine speed of marine diesel accordingly with the wind speed and its directions and also periodically to control the exhaust temperature and heat density of engine that to provide the normal process of its operation.

For instance, at wind speed 4 Beaufort ,we have the following corrections as shown in Figure 3:

- *At wind direction X_5 should not to correct primary the engine speed because this is the good conditions for the marine engine, i.e the ship moves with a tail wind (we accept the value of engine at this period about of* **$n= 110$ r/min as 100 percent);**

- *At wind direction X_1 ,i.e the ship moves against a headwind and this case should to decrease the primary engine speed on* **10 percent;**

- *At wind direction X_2 ,i.e the ship moves against a headwind which acts from the left side and in this situation should to decrease the primary its value on* **12 percent;**

- *At wind direction X_8 ,i.e the ship moves against a headwind which acts from the right side and in this situation should to decrease the primary its value on* **13.30 percent.**

As we see from the Figure 3, the most unfavorable conditions for the operation of marine diesel and ship in the tropics are the wind directions X_2, X_6 and X_4, X_8 because they demand of decreasing of engine speed and besides they create the twisting moment, i.e the ship at this moment " turn around " and operational people at this time continue to keep the primary engine speed.

These actions made the operation of marine engine more difficult because at this time the exhaust temperature and heat density very high.

The results of evaluation regression models for engine speed against wind speed and its directions are shown in Table 4.

Table 4 ANOVA table for wave sea(wind speed and its directions)-engine speed in the different regression models

Direction wind	Regression model	Sum of squares		Degree of freedom	Mean square		Coefficient of determination	Coefficient of correlation
		SSR	SSE		MSR	MSE		
X_2	$\hat{Y}=103.55+1.04 X_2$ $i=9$	37.6	65.9	17	37.6	9.42	0.36	0.60
X_3	$\hat{Y}=103.21+0.79 X_3$ $i=6$	5.8	61.1	14	5.8	15.3	0.09	0.29
X_4	$\hat{Y}=84.83+3.46X_4$ $i=4$	104.6	198.2	12	104.6	99.1	0.35	0.59
X_5	$\hat{Y}=104.8+1.2X_5$ $i=6$	4.80	37.20	14	4.80	9.30	0.11	0.34
X_6	$\hat{Y}=104+X_6$ $i=5$	7.20	20	13	7.2	6.70	0.26	0.51
X_7	$\hat{Y}=98.37+1.5X_7$ $i=9$	65.4	16.9	17	65.4	2.41	0.79	**0.89**
X_8	$\hat{Y}=103.05+0.83X_8$ $i=12$	36.30	81.50	110	36.30	8.15	0.31	0.56
X_1	$\hat{Y}=101.59+1.97X_1$ $i=41$	161	838	140	161	20.95	0.16	0.40

The change in engine speed with wind direction and average wind speed ($\overline{W_s}=3$) is shown in Figure 4.

This regression function $Y_4=\gamma(X_4)$ has the linear model and describes by the equation $\hat{Y}_4=106.81-0.01X_4$.

Figure 4 also shows the change in engine speed for the different wind directions. The value of engine speed, as we see in Figure 4,decreases particularly with the wind direction X_1 ,i.e when the ship moves against a headwind.

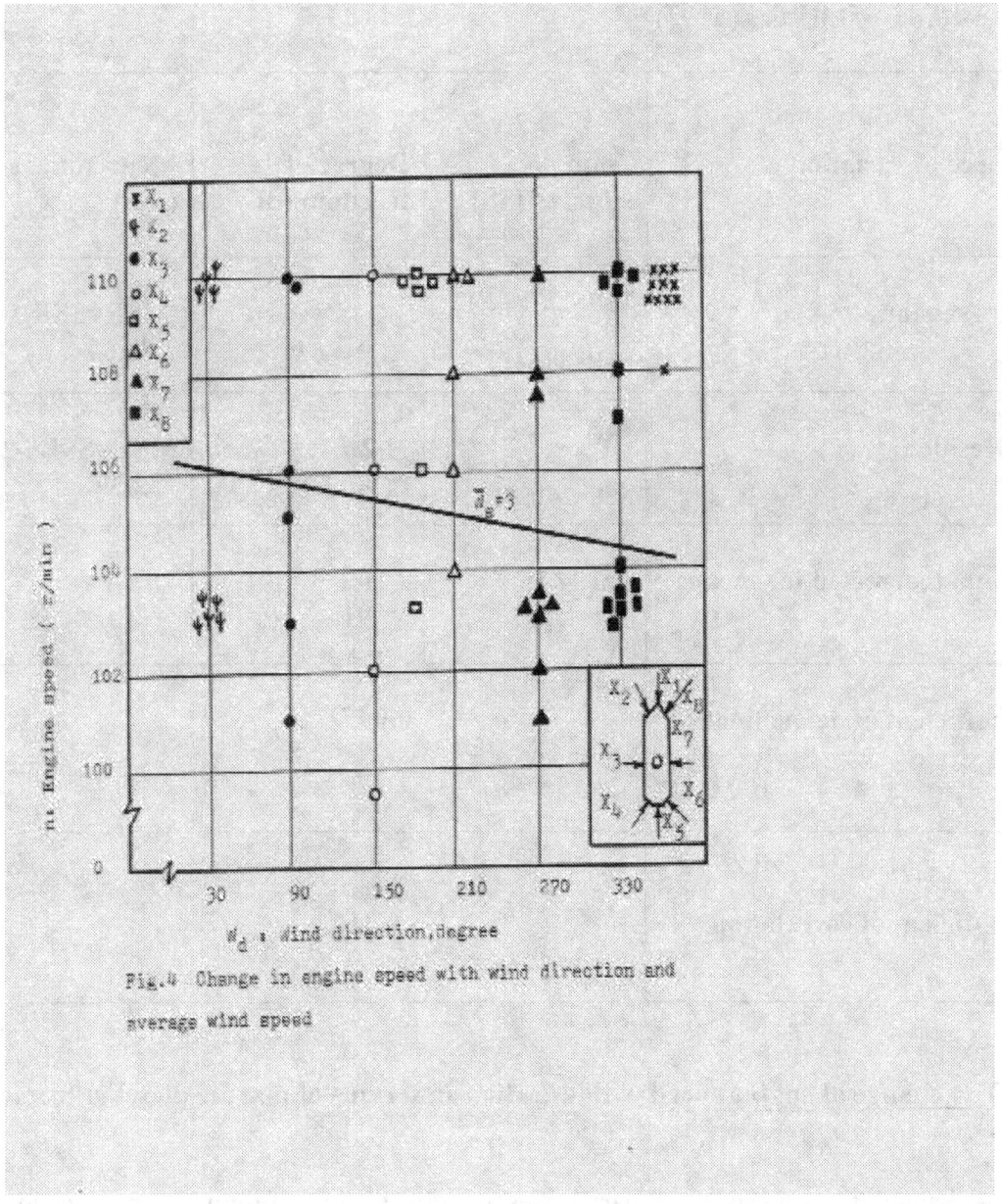

Fig.4 Change in engine speed with wind direction and average wind speed

These results confirm the above-named conclusions that at this situation it is necessary to decrease the engine speed so that to guarantee the normal heat density and exhaust temperature of marine diesel engine in the tropics.

In the Table 5 is shown the ANOVA table for the wind direction-engine speed for

regression model $\hat{Y} = 106.81 - 0.01 X_4$ (i=92) .

31

Anatoly Rozenblat

Table 5 ANOVA table for the wind direction-engine speed for regression model

$$\hat{Y} = 106.81 - 0.01\ X_4'\ (\ i=92)$$

Source of variation	Sum of squares (SS)	Degree of freedom (df)	Mean square (MS)
Regression	$SSR=\sum(\hat{Y}-\bar{Y})^2$ SSR=88.64	1	MSR=SSR/1 88.64
Residual	$SSE=\sum(Y-\hat{Y})^2$ SSE=2015.51	i-2 90	MSE=SSE/i-2 22.39
Total (corrected for mean)	2104.15	i-1 91	
Coefficient of determination	0.04		
Coefficient of correlation	- 0.21		

2.5 The change of engine speed with duration-in service of marine diesel in tropics

The change of engine speed of marine diesel engine against of duration-in service is shown in Figure 5.

From the Figure 5 we see that the functional analysis $Y_2 = \psi\ (\ X_2)$ has the second-order polynomial regression model for one predictor variable which presents in view of parabola and describes by the equation $\hat{Y} = 149.27 - 3.13\ X_2' + 0.02\ X_2'^2$.

And besides the Figure 5 also shows that engine speed decreases with the increasing of duration-in service of ship (diesel engine0 in the tropics.

The decreasing of engine speed could be explained that body ship as the fouling allowance. It is seen from Figure 5 the scatter plot (observed data) distributes on three zones:

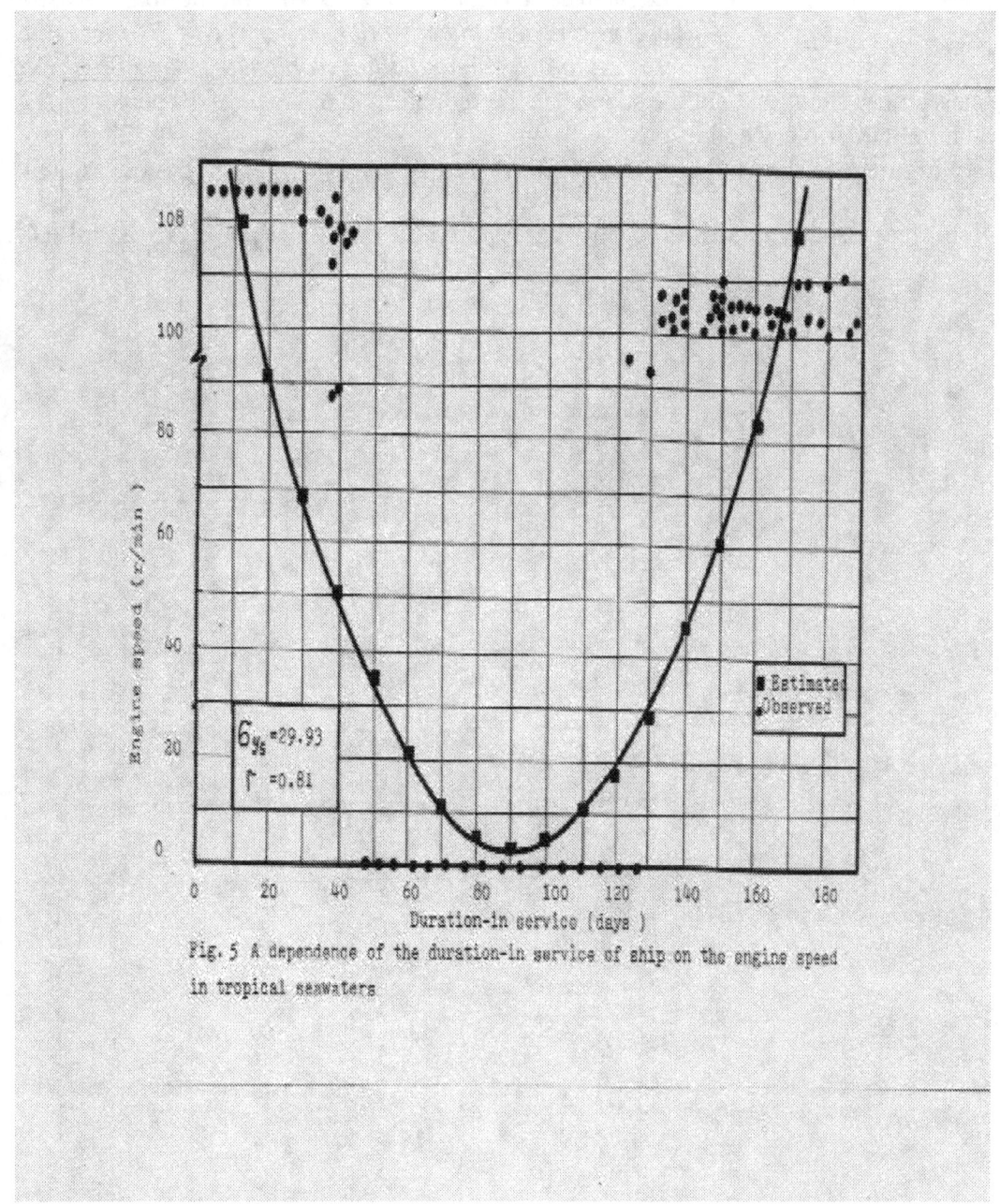

Fig. 5 A dependence of the duration-in service of ship on the engine speed in tropical seawaters

- *The first zone shows the motion of ship in primary conditions (the fouling is equal zero) : running area —Black Sea to Indian Ocean and other tropical seas with duration-in service of 43 days;*

- *The second zone shows the conditions when the ship and marine diesel engine without of operation with duration-in service of 85 days in the tropics;*
- *The third zone relates to the operational period of marine diesel engine with the duration-in service of 50 days: running area- Indian Ocean and other tropical seas and Black Sea.*

These zones show that engine speed has the reduction in average of 10 percent relatively to primary conditions.

In Figure 6 is shown the graphical analysis for several different types of residual plots engine speed versus X_2 for the regression model $Y' = 149.27 - 3.13\,X_2' + 0.02X_2'^2$

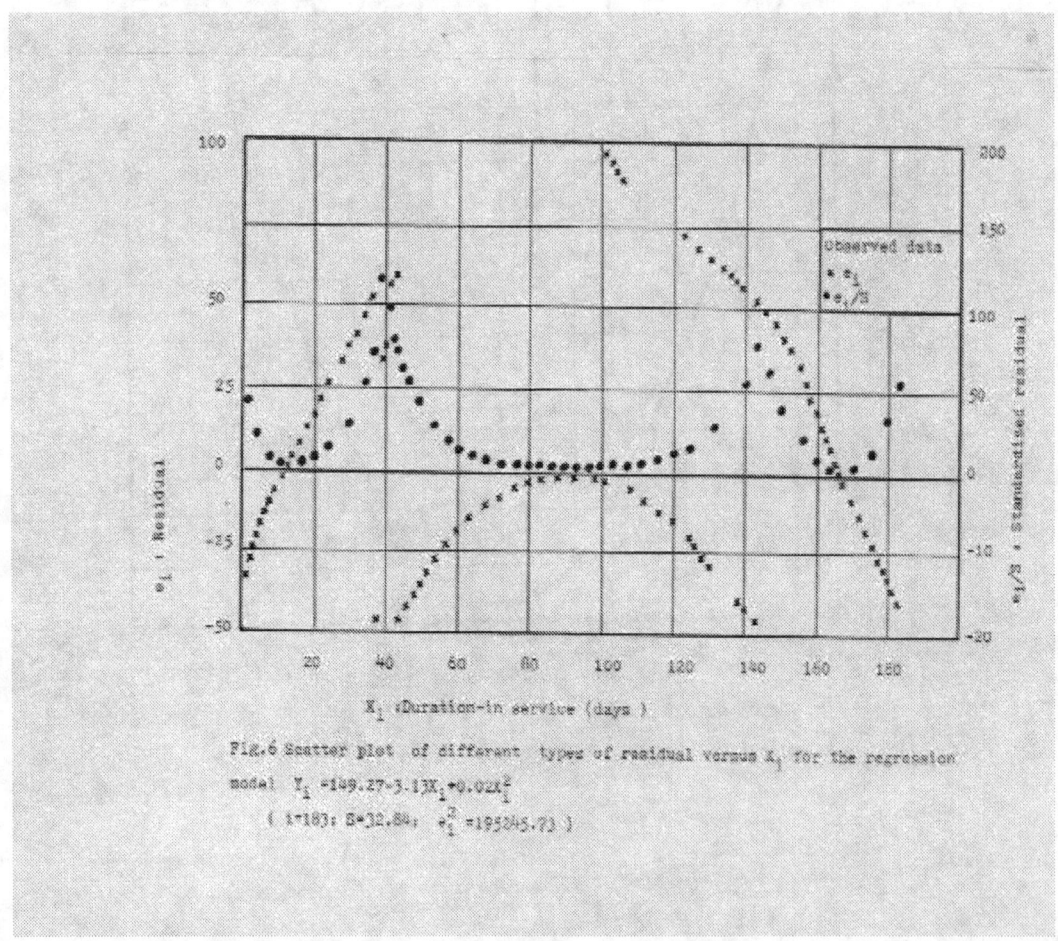

Fig.6 Scatter plot of different types of residual versus X_1 for the regression model $Y_1 = 149.27 - 3.13X_1 + 0.02X_1^2$

(i=183; S=12.84; e_1^2 =195845.73)

2.6 Multiple regression analysis in evaluation of engine speed

Functionally the dependence of engine speed of marine diesel engine in the multiple regression analysis has view :

$$Y = \beta_0 + \beta_1 X_1' + \beta_2 X_2' + \beta_3 X_3' + \beta_4 X_4' \qquad (1)$$

where β_0 ; β_1 ; β_2; β_3; and β_4 the coefficients of this model.

These coefficients could be determined from the system of equation in view of :

$$
\begin{pmatrix}
n & \sum X_{1,i} & \sum X_{2,i} & \sum X_{3,i} & \sum X_{4,i} \\
\sum X_{1,i} & \sum X_{1,i}^2 & \sum X_{1,i}X_{2,i} & \sum X_{1,i}X_{3,i} & \sum X_{1,i}X_{4,i} \\
\sum X_{2,i} & \sum X_{2,i}X_{1,i} & \sum X_{2,i}^2 & \sum X_{2,i}X_{3,i} & \sum X_{2,i}X_{4,i} \\
\sum X_{3,i} & \sum X_{3,i}X_{1,i} & \sum X_{3,i}X_{2,i} & \sum X_{3,i}^2 & \sum X_{3,i}X_{4,i} \\
\sum X_{4,i} & \sum X_{4,i}X_{1,i} & \sum X_{4,i}X_{2,i} & \sum X_{4,i}X_{3,i} & \sum X_{4,i}^2
\end{pmatrix}
\begin{pmatrix}
\beta_0 \\ \beta_1 \\ \beta_2 \\ \beta_3 \\ \beta_4
\end{pmatrix}
=
\begin{pmatrix}
\sum Y_i \\ \sum X_{1,i}Y_i \\ \sum X_{2,i}Y_i \\ \sum X_{3,i}Y_i \\ \sum X_{4,i}Y_i
\end{pmatrix}
$$

$$\sum Y_i = n\beta_0 + \beta_1\sum X_{1,i} + \beta_2\sum X_{2,i} + \beta_3\sum X_{3,i} + \beta_4\sum X_{4,i}$$

$$\sum X_{1,i}Y_i = \beta_0\sum X_{1,i} + \beta_1\sum X_{1,i}^2 + \beta_2\sum X_{1,i}X_{2,i} + \beta_3\sum X_{1,i}X_{3,i} + \beta_4\sum X_{1,i}X_{4,i}$$

$$\sum X_{2,i}Y_i = \beta_0\sum X_{2,i} + \beta_1\sum X_{2,i}X_{1,i} + \beta_2\sum X_{2,i}^2 + \beta_3\sum X_{2,i}X_{3,i} + \beta_4\sum X_{2,i}X_{4,i}$$

$$\sum X_{3,i}Y_i = \beta_0\sum X_{3,i} + \beta_1\sum X_{3,i}X_{1,i} + \beta_2\sum X_{3,i}X_{2,i} + \beta_3\sum X_{3,i}^2 + \beta_4\sum X_{3,i}X_{4,i}$$

$$\sum X_{4,i}Y_i = \beta_0\sum X_{4,i} + \beta_1\sum X_{4,i}X_{1,i} + \beta_2\sum X_{4,i}X_{2,i} + \beta_3\sum X_{4,i}X_{3,i} + \beta_4\sum X_{4,i}^2$$

At the following data which are shown in Table 6 we could later to calculate the coefficients β_0; β_1; β_2; β_3 and β_4.

Table 6 Calculation of coefficients

n=183	$\sum Y = 9678$	$\sum X_1 = 4649.50$	$\sum X_2 = 16836$	$\sum X_3 = 240$

$\sum X_4 = 24690$	$\sum X_1 Y = 220561.5$	$\sum X_1^2 = 120901.25$	$\sum X_1 X_2$ $=431498.50$	$\sum X_1 X_3 = 5249$
$\sum X_1 X_4 = 576585$	$\sum X_2 Y = 902327$	$\sum X_2^2 = 2059604$	$\sum X_2 X_3 = 17591$	$\sum X_2 X_4 = 2407350$
$\sum X_3 Y = 25545$	$\sum X_3^2 = 940$	$\sum X_3 X_4 = 57510$	$\sum X_4 Y = 2590200$	$\sum X_4^2 = 7838100$
$\beta_0 = 76.338$	$\beta_1 = -2.739$	$\beta_2 = 0.076$	$\beta_3 = 9.379$	$\beta_4 = 0.199$

So, the multiple regression model expresses in view of equation:

$$\hat{Y}_i = 76.338 - 2.739 X_{1,i} + 0.076 X_{2,i} + 9.379 X_{3,i} + 0.199 X_{4,i} \qquad (3)$$

In Table 7 is shown ANOVA table for engine speed (Y) regressed on seawater temperature (X_1) ,duration-in service (X_2) ,wave sea (wind speed (X_3) and wind direction (X_4).

Table 7   ANOVA table for multiple regression model

$$Y_i = 76.338 - 2.739 X_1 + 0.076 X_2 + 9.379 X_3 + 0.199 X_4$$

Source	Degree of freedom (df)	Sum of squares (SS)	Mean square (MS)	Variance ratio (F)	Coefficient correlation (R)
Regression	4	SSR=445953.76	111488.40	328.61	
Residual	181	SSE=61409.12	339.28		
Total	185	SSY= 507362.88			0.938

Conclusions

36

1.*Multiple regression model for evaluation of engine speed of two-stroke cycle marine engine from the different external factors such as : seawater temperature ,wave sea (wind speed and it directions) and duration-in service in the tropics is developed in this work;*

2.*The result which have been obtained in these regression models differ from the present recommendations* **[Osbourne A., Modern Marine Engineer's Manual,1944 ,vol.11 ,pp.15-1219-***Cornell Maritime Press ,New York]* *which indicate that engine speed is only the component of engine horsepower;*

3.*Operation of two-stroke cycle diesel engine in the tropics shows that engine speed decreases and this value will be greater with increasing of duration-in service of ship in the tropical seawaters. For these conditions the average dropping engine speed is the 10 percent;*

4. *However, the dropping of engine speed has a place also in wave sea conditions. In this situation should to reduce the engine speed in dependence from the wind speed and its direction;*

5.*Regression model in dependence of engine speed against duration-in service has view of polynomial and describes as parabola;*

6. *Regression model in dependence of engine speed from wave sea (wind speed and its directions) and seawater temperature have the linear function.*

CHAPTER 3

Statistical Analysis of Ship Speed in the Tropics

3.1 The advantages of statistical methods in the evaluation of the ship speed

The statistical methods used vary widely in the evaluation of the different problems in marine engineering by the author **(Perakis,1991) and (Zvonkov,1956).** However, they were used in regard to the questions of studying ship speed, possibly to relate only the information of the data of which was investigated on the basis of using a merchant vehicle in the conditions of the Caspian and Baltic Seas.

These papers also mark that the reduction of ship speed is the function of a duration in-service of a ship and this dependence has a nonlinear character. The conclusion of the author does not consider the questions of the ship reduction and does not take into account the ship standing in the ports or on the roadstead.

Some attempts of analysis of the ship speed were made by the authors **(Parker,1971) and (Lacey and Edwards,1993)**,but their concerns are mainly the conditions of sailing ships in cold seawaters.

Considering the above-named factors, the author of this paper thinks that the questions of the ship speed reduction in the conditions of a sailing ship in the tropics are of considerable interest for the seaman and navigational services.

The most important factor in the evaluation of total cost of transportation of commercial cargoes by sea is the ship speed **(Avallone and Baumenstein,1987).** However, this index is the complex function which depends on multiple independent variables and submits to the law of occasional values.

Naturally receiving reliable statistical data on the real object- merchant vehicle is of the most value and most important to the researcher because these data do not demand considerable capital investment on the realization or expensive laboratory investigation ,for example, testing in a special pool ,and besides , everything this will increase in the future of the range of using statistical methods in industrial conditions.

On the basis of the above-named circumstances, the author in this paper thinks that the correlation analysis, as one of the varieties of the statistical method in the evaluation of the ship ,is the optimum and the most rational.

The general objective of statistical analysis in this paper was to evaluate the ship speed in dependency of such areas as service diesel engine (engine speed, exhaust gas temperature, etc.), voyage duration of merchant vehicle , seawater temperature, and seaway, taking into account the sailing ship in the tropics.

The author, on the basis of the statistical data with the number of observation *I=183* attempts to determine the relationship in view of linear and nonlinear correlation between two variables. In the case of examining an independent variable, one of the parameters of this dependence is used as :

- X_1= *wind direction on the east (E),degree;*

- X_2= *wind direction on northeast (NE) ,degree;*

- X_3=*wind direction on north (N) ,degree;*

- X_4=*wind direction on northwest (NW),degree;*

- X_5=*wind direction on west (W),degree;*

- X_6=*wind direction on southwest(SW),degree;*

- X_7=*wind direction on south (S),degree;*

- X_8=*wind direction on southeast(SE),degree;*

- X_9=*revolution per minute of the main engine;*

- X_{10}=*duration in –service of ship , days;*

- X_{11}=*seawater temperature, $°C$;*

- X_{12}=*wind speed(Beaufort wind scale);*

- X_{13}=*exhaust gas temperature , $°C$;*

and dependent variable was chosen the value as: *Y= ship speed, miles per hour.*

So,the author in this paper investigates the following correlation dependencies such as :

$$Y_1 = \varphi_1(X_1);$$
$$Y_2 = \varphi_2(X_2);$$
$$Y_3 = \varphi_3(X_3);$$
$$\cdots\cdots\cdots\cdots$$
$$Y_{13} = \varphi_{13}(X_{13}).$$

which will be discussed in the section titled " Discussion and Results". This portion takes into account the peculiarities of the sailing merchant vehicle in the tropics.

3.2 Presentation and analysis of the statistical data

The analysis of sailing ship in the tropics in comparison with the other conditions of sailing has some peculiarities such as :

a) *For this ship one observes the intensive marine growth (fouling) in the period of voyage and accordingly this factor decreases the quick-maneuvering qualities of the ship as a whole;*

b) *The main marine diesel engine works in more critical conditions (very often have the place overheat and overload of diesel) ,and naturally these factors decrease the service life of the marine diesel engine as a whole;*

c) *The total price of transportation of cargo increases as a result of ship speed reduction ,which is joined with the duration of service for a ship, i.e with an increase in quantity of the mooring in ports and duration of service ,the ship speed decreases considerably;*

d) *For some objective reasons (such as the sailing ship in the tropics and the presence of the seaway) ,the unfavorable conditions on the examined object arise –of the ship ,main marine diesel engine ,crew ,cargo –and we see that all of these component parts are joined very tightly and demand the complex investigation.*

Using the data which are given in **Appendix 1,** we see in Figure 1.1 that the correlation between ship speed and engine speed as shown in view of function $Y_9=\varphi_9(X_9)$ has the linear relationship with regression line $V= 0.07+ 0.13n$ **(1)** with the following summary statistical characteristics:

Average engine speed	$\bar{n}=52.89$ *revolutions per minute;*
Average ship speed	$\bar{v}= 6.95$ *miles per hour;*
Coefficient correlation	$r=0.992;$
Variance	$S_{v/n}=1.29$

Figure 1.1 shows also that with increasing of engine speed the ship speed increases accordingly, i.e this function has the directly proportional dependence.

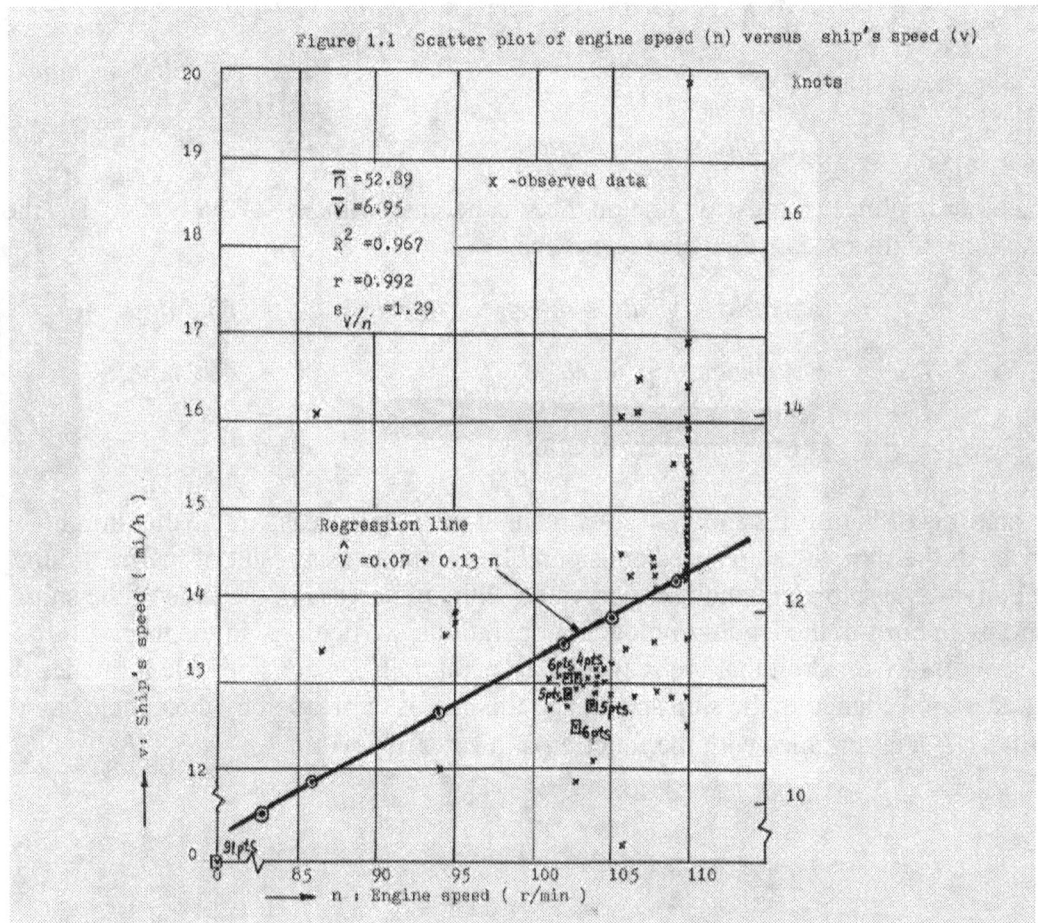

Figure 1.1 Scatter plot of engine speed (n) versus ship's speed (v)

However, a scatter- plot diagram of the duration in-service versus ship speed shown in Figure 1.2 indicates that the correlation between ship speed and duration in-service of a ship shown in view of function $Y_{10} = \varphi_{10}(X_{10})$ has the nonlinear relationship with two curves and regression functions such as :

$$\hat{V}_1 = 16.50\,N^{-0.0297} \qquad (2) \quad \text{and} \quad \hat{V}_2 = 17.36\,N^{-0.0568} \qquad (3)$$

The first curve with the function $\hat{V}_1 = 16.15\,N^{-0.0297}$ characterizes the first period of the sailing ship(without stopping in ports).And the second curve with the function

$\hat{V}_2 = 17.36\,N^{-0.0568}$ characterizes also the second period of the sailing ship (but already with the ship stopping in ports).

As shown in Figure 1.2 ,the duration of a ship standing in ports with the tropical climate for more than eighty days and the ship speed at this occasion accordingly decreases in comparison with the first pass of a ship (without of stopping).

The nonlinear regression line on the first pass with $\hat{V}_1 = 16.15\,N^{-0.0297}$ has the following summary statistical characteristics:

Average duration in-service of ship $\overline{N_1}=21\ days;$

Average ship speed $\overline{V_1}=14.89\ miles\ per\ hour;$
Variance $\sigma_1=0.038;$
Coefficient correlation $r_1=0.28$

The other nonlinear regression line on the second pass with $\hat{V_2}=17.36\ N^{-0.0568}$ has the following summary statistical characteristics:

Average duration in-service of ship $\overline{N_2}=163.5\ days\ ;$

Average ship speed $\overline{V_2}=12.99\ mph;$
Variance $\sigma_2=0.006;$
Coefficient correlation $r_2=0.25.$

So, analysis of Figure 1.2 shows as a whole that with an increase in the duration in-service of the ship, the ship speed considerably decreases as a result of marine fouling on the body of the ship and which is aggravated with an increase in the time of the ship standing in ports at the loading-unloading operations, particularly in the tropics .

It is necessary to admit that the seawater temperature has a considerable influence on the character of a change in the ship speed, and this fact is confirmed by the conclusion of Figure 1.2 in accordance with the function of $Y_{11}=\varphi_{11}(X_{11}).$

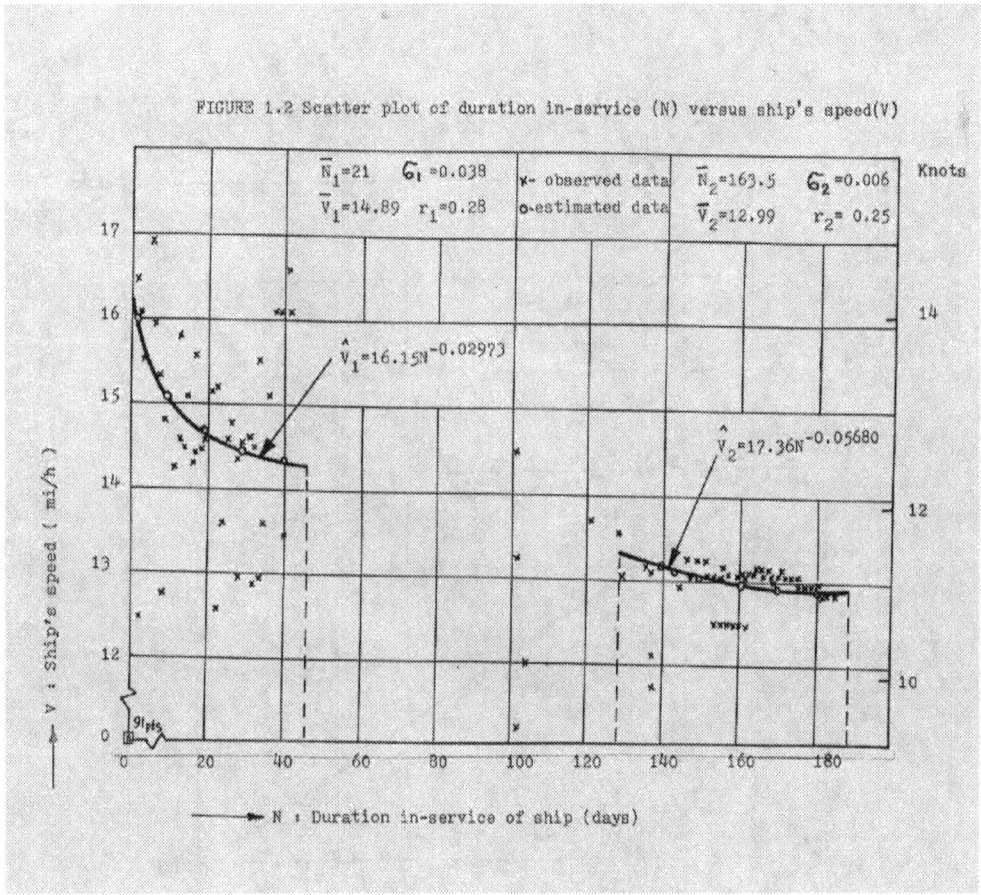

FIGURE 1.2 Scatter plot of duration in-service (N) versus ship's speed(V)

However, as Figure 1.3 shows ,the value of changes in the ship speed depends also on the character of the location of the ship in seawater.

At the mooring of a ship in the tropics (shown by a continuous line) the ship speed has the linear regression line $\hat{V}_1 = 41 - 1.34\, T_w$ (4) with the following summary statistical characteristics:

Average seawater temperature	$T_{w,1} = 25.41\ ^{\circ}C$;
Average ship speed	$\hat{V}_1 = 6.95$ *miles per hour*;
Coefficient correlation	$r_1 = -0.67$;
Coefficient of determination	$R_1^2 = 0.76$.

And at the absence of mooring for a ship in the tropics (as shown by the dash line) the ship has the linear regression line $\hat{V}_2 = 15.61 - 0.04\, T_w$ (5).

So, analysis of Figure 1.3 shows that ,with increasing of seawater temperature ,the ship speed abruptly decreases ,particularly with an increase in the mooring of a ship in ports or on the roadstead.

43

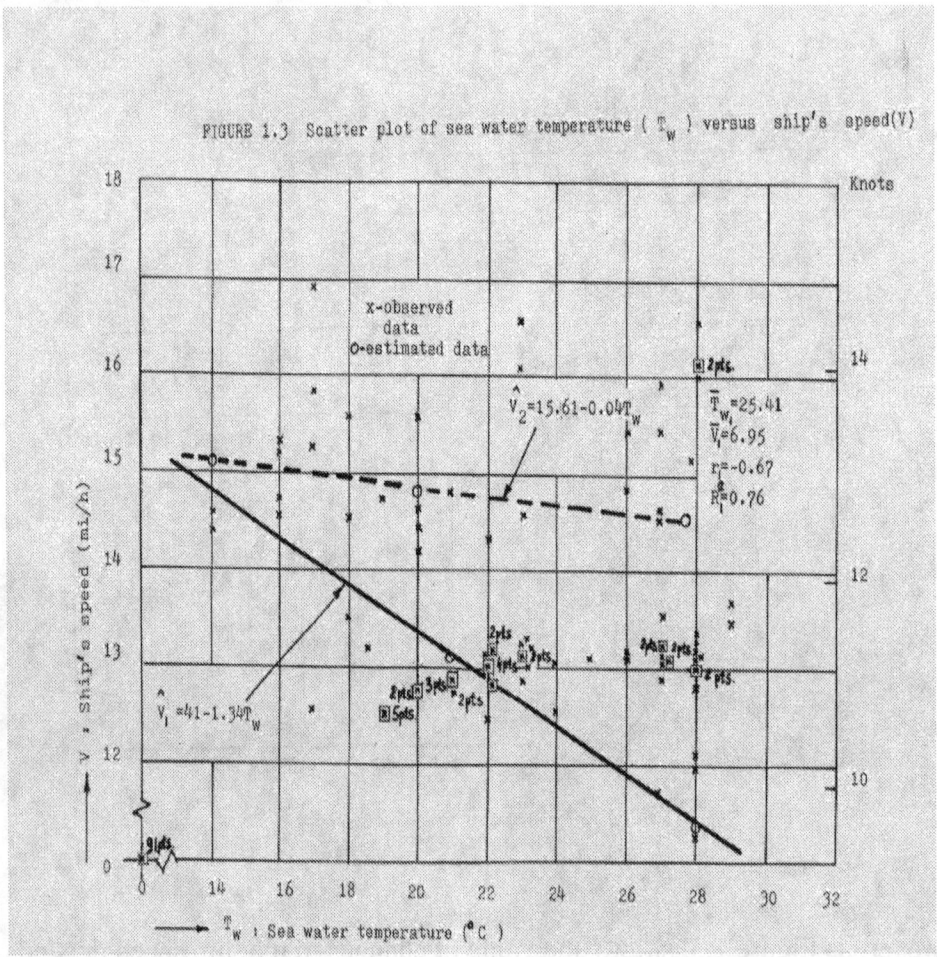

FIGURE 1.3 Scatter plot of sea water temperature (T_w) versus ship's speed(V)

No less important question in analysis of ship speed in the tropics, besides the above-named factors, is the seaway factor. So, the data indicated in Figure 4 shows that, besides the effects of wind speed on the ship speed ,the direction of wind, which acts directly on the body of ship in the period of its motion ,also has a considerable influence.

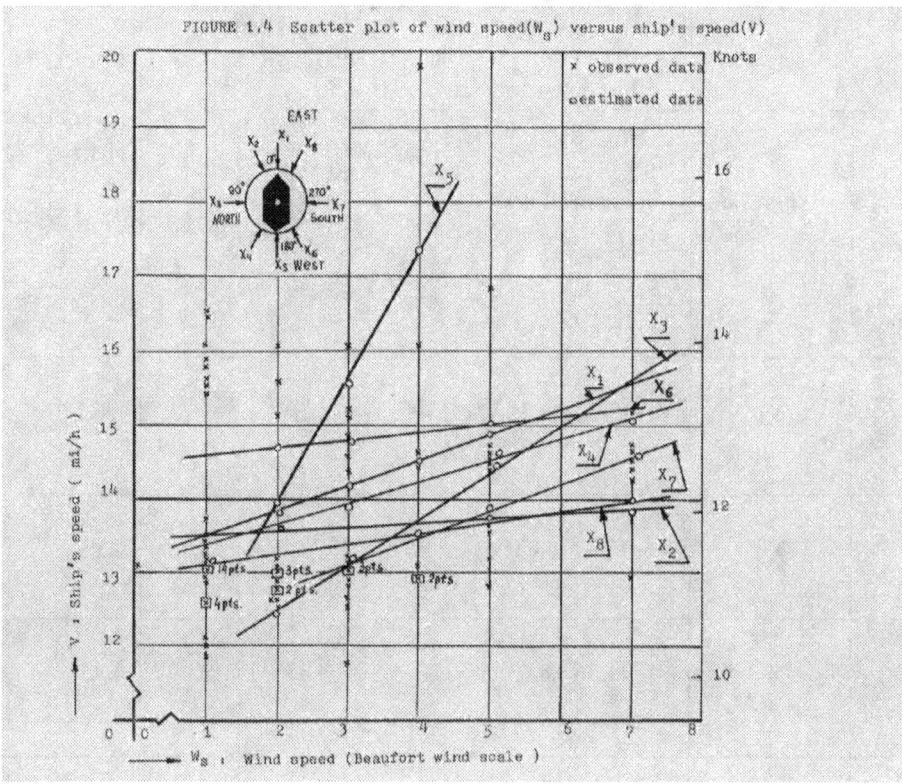

FIGURE 1.4 Scatter plot of wind speed(W_s) versus ship's speed(V)

From Figure 1.4 we see that the changing ship speed has the linear regression analysis for any direction of wind with the variable parameters of wind speed.

In Table 1 are shown the summary characteristics of the wind speed (W_s) versus ship's speed (V) and evaluated for each view of the direction of the wind , having the linear regression lines.

As shown in Figure 1.4 with the direction (X_5) of wind to the stern of the ship, the ship's speed considerably increases and then accordingly is larger than the value of wind speed (W_s);that larger number is the ship's speed.

However, the reduction of ship speed has a place also at the direction of wind (X_2) and (X_8) ,and the ship's speed particularly decreases with the rising of wind speed.

So, the presence of direction (X_5) of the wind to the stern of the ship promotes the acceleration of motion in the ship even at the presence of fouling on the body of a ship in the tropics, and besides ,the above-named conditions considerably improve the regime of service of the main diesel engine.

Table 1 Summary characteristics of the wind speed (W_s) versus ship's speed (V)

Wind directions	Estimated regression	Statistical characteristics

degree		line		Arithmetic mean		*Correla-tion* coeffi-cient r	*Coeffi-cient of* determ i-nation R^2
Name	Designation	Ship speed (V_i)	*Ob-served* data	Wind Speed (Ws)	Ship's speed		
East (0)	X_1	$V_1=13.23 +0.35$ Ws	41	1.39	13.71	0.28	0.28
Northeast (30)	X_2	$V_2=13.64+0.03$Ws	9	3.11	13.73	0.13	0.05
North (90)	X_3	$V_3=11.36 +0.63$ Ws	6	3.33	13.42	0.93	0.93
Northwest (150)	X_4	$V_4=13.04+0.297$Ws	4	4.75	14.45	0.35	0.35
West (180)	X_5	$V_5=10.21+1.81$Ws	6	2.67	15.05	0.51	0.51
Southwest (210)	X_6	$V_6=14.69+0.05$Ws	5	3.60	14.87	0.04	0.04
South (270)	X_7	$V_7=12.02+0.35$Ws	9	4.11	13.44	0.83	0.83
Southeast (330)	X_8	$V_8=12.83+0.18$Ws	12	3.75	13.50	0.49	0.43

Figure 1.5 shows the scatter plot of exhaust gas temperature versus ship's speed. Exhaust gas temperature indirectly influences on the ship speed, and from Figure 1.5 we see that the changing of this functional dependence $Y_{13}=\varphi(X_{13})$ has the linear regression line

$\hat{V} = 18.799 - 0.012\, T_g$ (6) with the following summary statistical parameters:

Average engine speed $\bar{n}=92$ *revolutions per minute;*

Average exhaust gas temperature $\bar{T_g}=412.47$ °C;

Average ship speed $\bar{V}=13.83$ *miles per hour* ;

Coefficient correlation $r=-0.23;$

Coefficient of determination $R^2=0.20.$

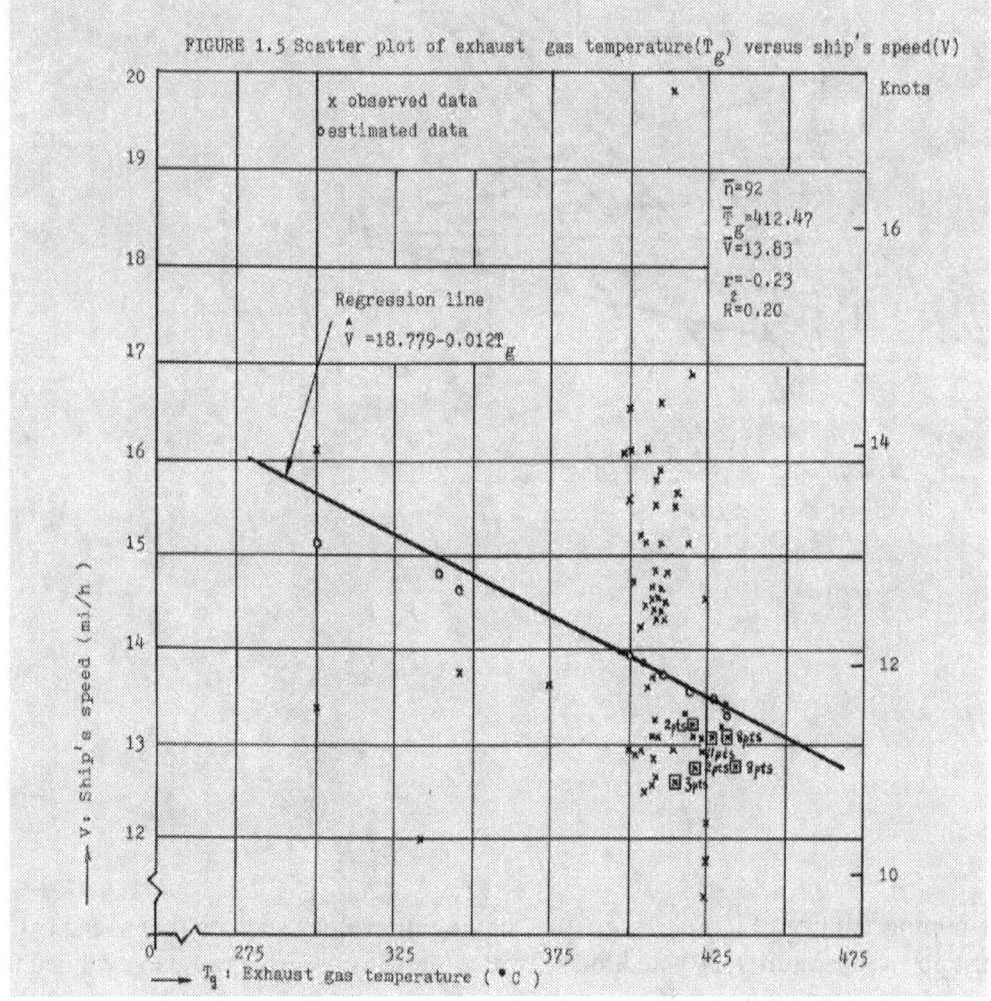

FIGURE 1.5 Scatter plot of exhaust gas temperature(T_g) versus ship's speed(V)

From Figure 1.5 we see that with the decreasing exhaust gas temperature, i.e the engine speed, the ship speed considerably decreases.

Analyzing the multiple regression analysis for the ship speed in dependence on some such variables as **(X9), (X11) and (X13)** we see from Figure 1.6,taken as the base for the calculation of nomogram ,that the ship's speed in the tropics has the linear regression line and is expressed by the formula :

$$\hat{V_c} = -0.7803 + 0.233\ X_9 - 0.0256\ X_{13} + 0.0283\ X_{11} \qquad (7)\ or$$

$$\hat{V_c} = -0.7803 + 0.233\ n - 0.0256\ T_g + 0/0283\ t_w \qquad (7\ a)$$

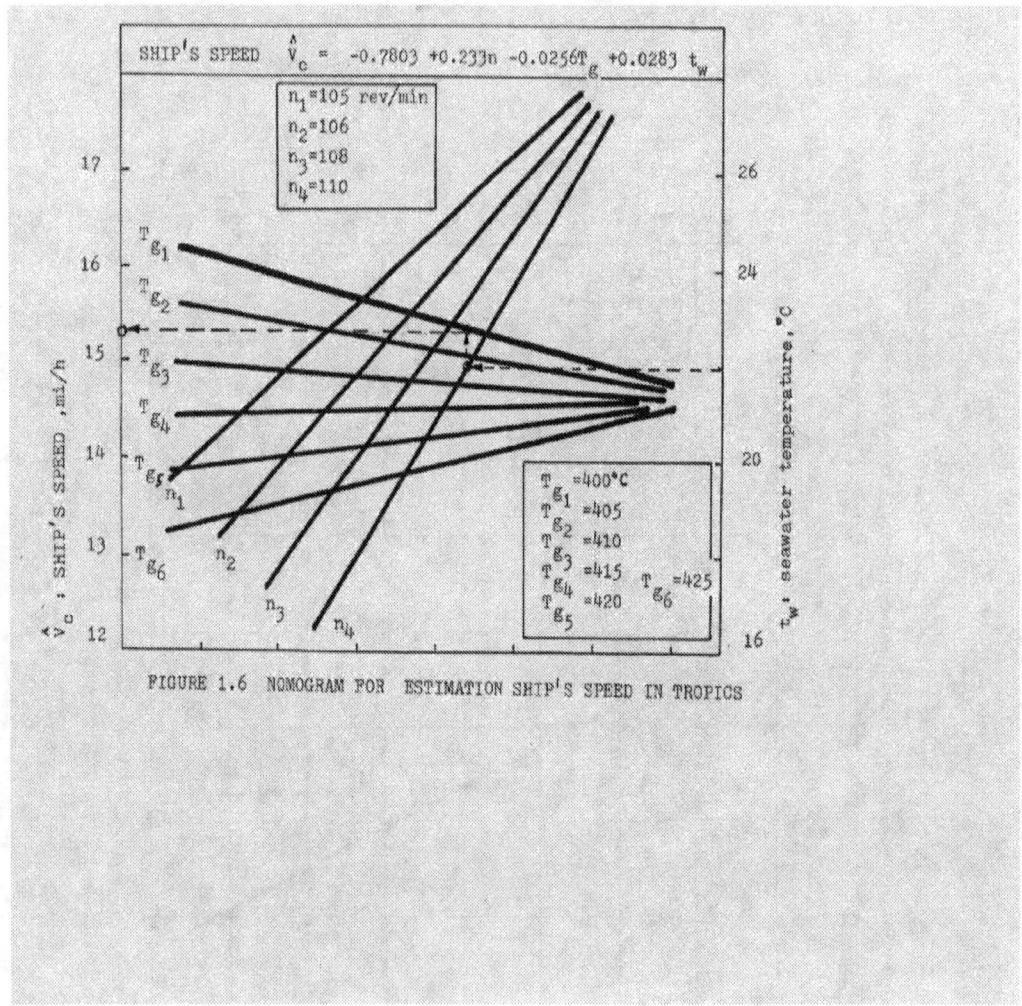

FIGURE 1.6 NOMOGRAM FOR ESTIMATION SHIP'S SPEED IN TROPICS

As the nomogram from Figure 1.6 testifies, having the values such engine speed (n) ,exhaust gas temperature (T_g) ,and the seawater temperature (t_w) ,we can evaluate the ship speed with precision to 0.50 percent.

So, at the data $t_w = 22\,°C$, $n_4 = 110\ rpm$ $T_{g,1} = 400\,°C$ from the nomogram in Figure 1.6 we have $V_c = 15.30\ mph$. The ship speed calculated on the formula (7a) has the value $V_{c,1} = 15.23\ mph$, i.e the relative error is equal to 0.50 percent.

Table 2 shows the average characteristics of changing the parameters of ship speed

($\bar{V_c}$) , the seawater temperature (t_w) ,revolution per minute of the main engine (engine speed) $\bar{n}$,and exhaust gas temperature ($\bar{T_g}$) depending on duration of in-service time of a ship in the tropics.

Table 2 Average observed data for the merchant ship, sailing in the tropics

Duration of service time of ship, months	Ship's speed V_c mi/h	Seawater t_w , °C	Engine speed n r/min	Exhaust gas temperature T_g °C
1	14.91	19.57	109.70	408.70
2	5.43	27.93	38.20	142.43
3	0	28.20	0	0
4	1.70	20.53	13.73	53.47
5	6.10	26.87	47.73	192.57
6	12.96	22.68	102.94	425.21

The data from Figure 1.7 and Table 2 shows that the most maximum seawater temperature has a place in the middle period of a ship sailing in the tropics (duration of in-service time of a ship is equal to 3, and the seawater temperature is equal to t_w=**28.20°C.**

Besides , the values $\bar{n}$,$\bar{T_g}$ and $\bar{V_c}$ at this period of a ship sailing in the tropics have the minimum values ($\bar{V_c}$=**0 ; n=0; T_g=0**),i.e the ship at this period was standing ,and naturally all these factors accelerated with the fouling on the body of the ship at this period.

On the basis of this data, the author makes the conclusion that after the long standing time of a ship in the tropics, all parameters of the ship, particularly the engine speed become worse, i.e the considerable reduction of the ship is observed for the next period of sailing in the tropics.

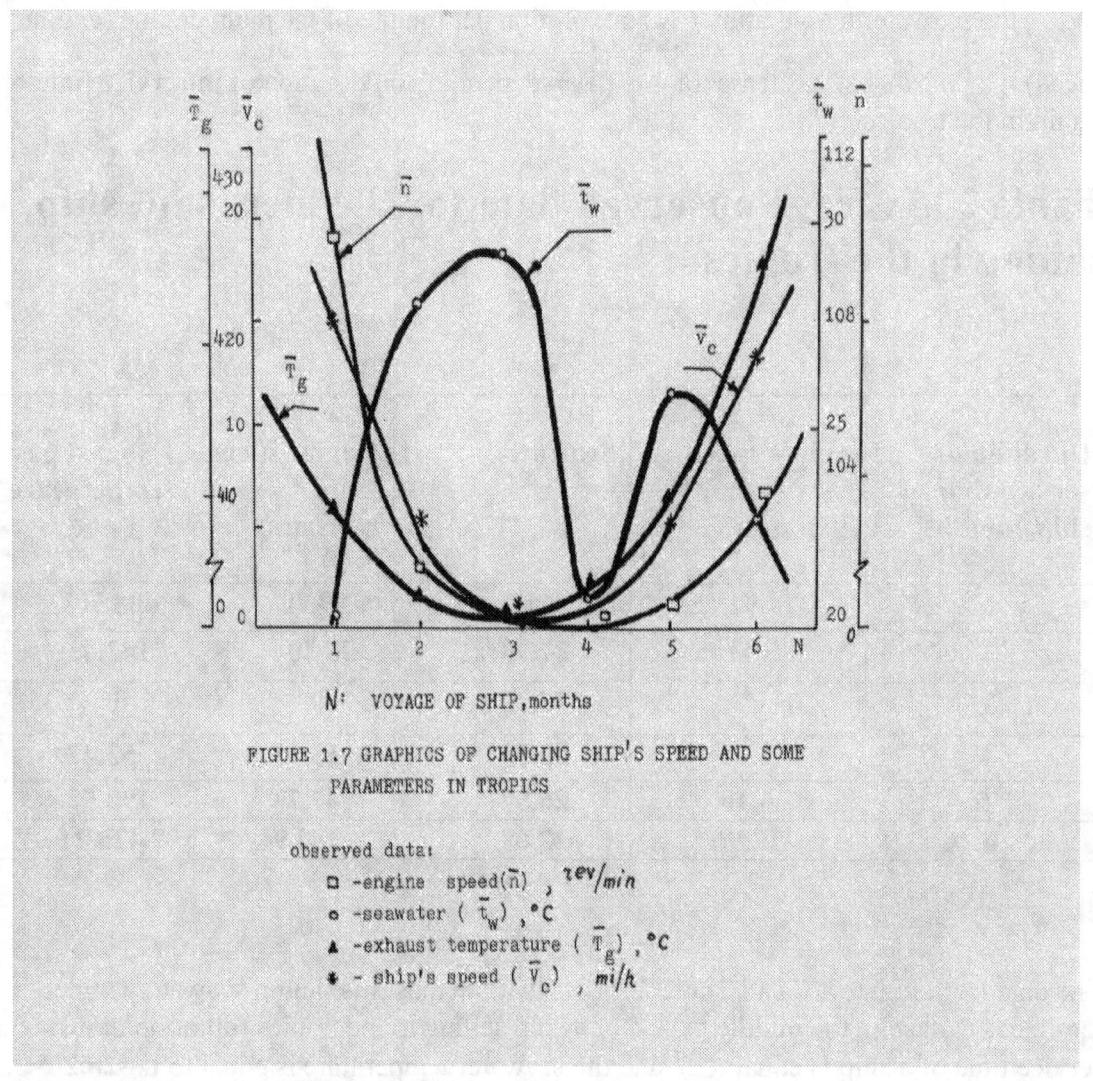

N: VOYAGE OF SHIP, months

FIGURE 1.7 GRAPHICS OF CHANGING SHIP'S SPEED AND SOME
PARAMETERS IN TROPICS

observed data:
- □ —engine speed($\bar{n}$) , rev/min
- ○ —seawater ($\bar{t}_w$) , °C
- ▲ —exhaust temperature ($\bar{T}_g$) , °C
- ✳ — ship's speed ($\bar{V}_c$) , mi/h

3.3 Conclusions and recommendations

Analyzing the reduction of ship speed for the merchant vehicle in the conditions of
sailing in the tropics ,the author of this paper makes the following conclusions:

a) *At the beginning of a ship's sailing in the tropics , the reduction of ship*
   speed observed average 30 percent in comparison to the starting speed;
b) *The ship speed is joined functionally to many variable values such as:*
   - *revolution per minute of the main engine (engine speed);*
   - *exhaust gas temperature;*
   - *seawater temperature;*
   - *wind speed;*

- *direction of wind;*
- *duration of in-service time of ship.*

But the most important factor in the question of reduction of ship speed is the duration of in-service time of a ship ,particularly in the tropics ,which also indirectly influences the other parameters of the main diesel engine (engine speed ,exhaust gas temperature ,etc.).

   c) *The reduction of ship speed is considerable for the tropics ,and this negative factor influences on the increasing of the total price of transportation for the commercial cargoes by the maritime way to many countries (India, New Zealand, Indonesia and other) with the same tropical climate;*

   d) *The correlation analysis of the change in the ship speed from some independent parameters such as engine speed ,seawater temperature ,exhaust gas temperature ,and parameters of the seaway (wind speed and its directions) shows that this dependence has the linear regression model;*

   e) *The correlation analysis of the change in the ship speed from an independent parameter such as the duration of in-service time of a ship has the nonlinear regression model;*

   f) *One index for decreasing the fouling on the body of the ship is reducing the quantity of mooring time for a ship in ports or on the roadstead particularly in the tropics;*

   g) *An increase in the exhaust gas temperature of the main diesel engine in a period of a ship sailing in the tropics is the first characteristic of the presence of fouling on the body of a ship.*

References

[**1**] Perakis Inozu," Statistical Analysis of Failure Time Distribution for Great Lakes Marine-Diesels Using Censored Data" , *Journal of Ship Research* ,(March ,1991):73.

[**2**] V.V.Zvonkov ,*" Marine Tow Tractor Calculations " (*Moscow:River Transport Publisher,1956):51-71.

[**3**] B.W.Parker," Marine Transportation in Alaska's Bering Sea and Arctic Ocean Areas," *Proceedings of the First International Conference on Port and Ocean Engineering under Arctic Conditions,(1971):1: 790-802.*

[**4**] P.Lacey and R.Edwards, "ARCO Tanker Slamming Study, "*Marine Technology, (July,1993): 135-147.*

[**5**] Eugene A. Avallone and Theodore Baumeister III ,*Mark's Standard Handbook for Mechanical Engineers,(New York: McGraw-Hill Book Company,1987):11-50.*

Bibliography

Croxton, Frederick E. and Cowden, Dudley J. *Applied General Statistics,* New York: Prentic-Hall,Inc.1939.

Kreyzig,Erwin, *Introduction to Mathematical Statistics* ,New York : John Willey&Sons,Inc.,1970

Morrison, Donald E. *Applied Linear Statistical Methods* ,Englewood Cliffs, New Jersey: Prentic-Hall ,Inc.,1983.

Pfaffenberger, Roger C. and Patterson, James H., *Statistical Methods,* Homewood, Illinois; Richard D.Irwin, Inc.,1977

CHAPTER 4

Functional Analysis of the Ship Speed Reduction in Waves and the Tropics and some Characteristics of the Main Engine

4.1 Some problems of motion ship in the tropical seawaters and the ways of its realizations

On the basis of ship in the Caspian and Baltic Seas as described by the author (Zvonkov,1956) ,**we see that the reduction of ship speed is the function of the duration of ship navigation and the conditions of sea waves.**

These conclusions, mainly questions regarding sea waves, were given by Lacey and Edwards in 1993 and Washio in 1994,**conforming to the North Pacific and Atlantic Oceans.**

In the other cold working area, such as the Antarctic Ocean and the Bering Sea these conclusions were marked by Parker and Vossers **which indicated that the wind and waves are at their most severe (especially in wintertime) and that these factors play a prevailing role over other factors in the reduction of ship speed .**

These problems in the above-named papers were not discussed conformably for the ship sailing in the tropical seas where the factor of duration of ship navigation had the larger role than sea waves in the questions of reduction of ship speed.

The general problem for the ships sailing in the tropical seawaters is the marine fouling of the body of the ship which significantly increases the coefficient of friction resistance and relative roughness.

Therefore, the total tractive resistance of the motion of the ship increases accordingly and ship's speed decreases, and these factors as a whole promote degradation of performance for the main engine.

This question was given great consideration by Crosby and Balasurbramanyan **only in regard to the prevention of marine fouling on the body of the ship. However, these papers did not examine deeply the functional analysis of the reasons of reduction on the ship and the work of the engine in the tropical seawaters.**

The author in this paper took as a basis some main objectives:

    a) *To discover the functional analysis of ship speed from the influence of the different factors;*

    b) *To evaluate and analyze the changes of ship speed and some parameters of the main engine by the methods of mathematical statistics on the basis of data from a dry-cargo ship with above –named described characteristics;*

c)   *To build histograms and frequency distributions for these parameters
and to describe its characteristics and shapes of distributions;*

d)   *To give some recommendations for research workers and navigators in
the questions of improving the work of the main engine and increasing
ship speed in the tropics.*

4.2 The functional analysis of the motion ship

**The author of this paper thinks that ship speed is the multiple of independent
variables such as :**

$$Y = f (X_1;X_2;X_3;\ldots\ldots\ldots X_{13}) \qquad (1)$$

where,

   Y = ship speed (dependent variable),miles per hour;

and independent variables such as :

- X_1= revolution per minute of the main engine;
- X_2=duration in-service time of ship, days;
- X_3 = temperature of seawater of outer cooling system of the main engine
  ,°C;
- X_4= wind speed (Beaufort wind scale);
- X_5=wind direction on the east (E) ,degree;
- X_6=wind direction on northeast (NE),degree;
- X_7=wind direction on southeast (SE),degree;
- X_8=wind direction on west (W) ,degree;
- X_9=wind direction on northeast (NW),degree;
- X_{10}=wind direction on southwest (SW),degree;
- X_{11}=wind direction on north (N) ,degree;
- X_{12}=wind direction on south (S) ,degree;
- X_{13}=temperature of exhaust gases from engine,°C.

Comparative analysis of the reduction of ship is shown in Figure 1 which proves
scientifically that in conditions of sailing a ship in the tropical seawaters, prevailing
factors of this reduction of ship speed is the temperature of seawater (X_3)and duration
of the in-service time of the ship (X_2) and sea wave directions $(X_5 \div X_{12})$ because these
factors promote an increase of the coefficient of tractive resistance and relative roughness
into account of the value of marine fouling on the body of the ship.

So, the author in this paper tries to analyze the influence of the different factors on variation of ship speed and to discover some characteristics of the work of the main engine of this ship, sailing in the tropical waters.

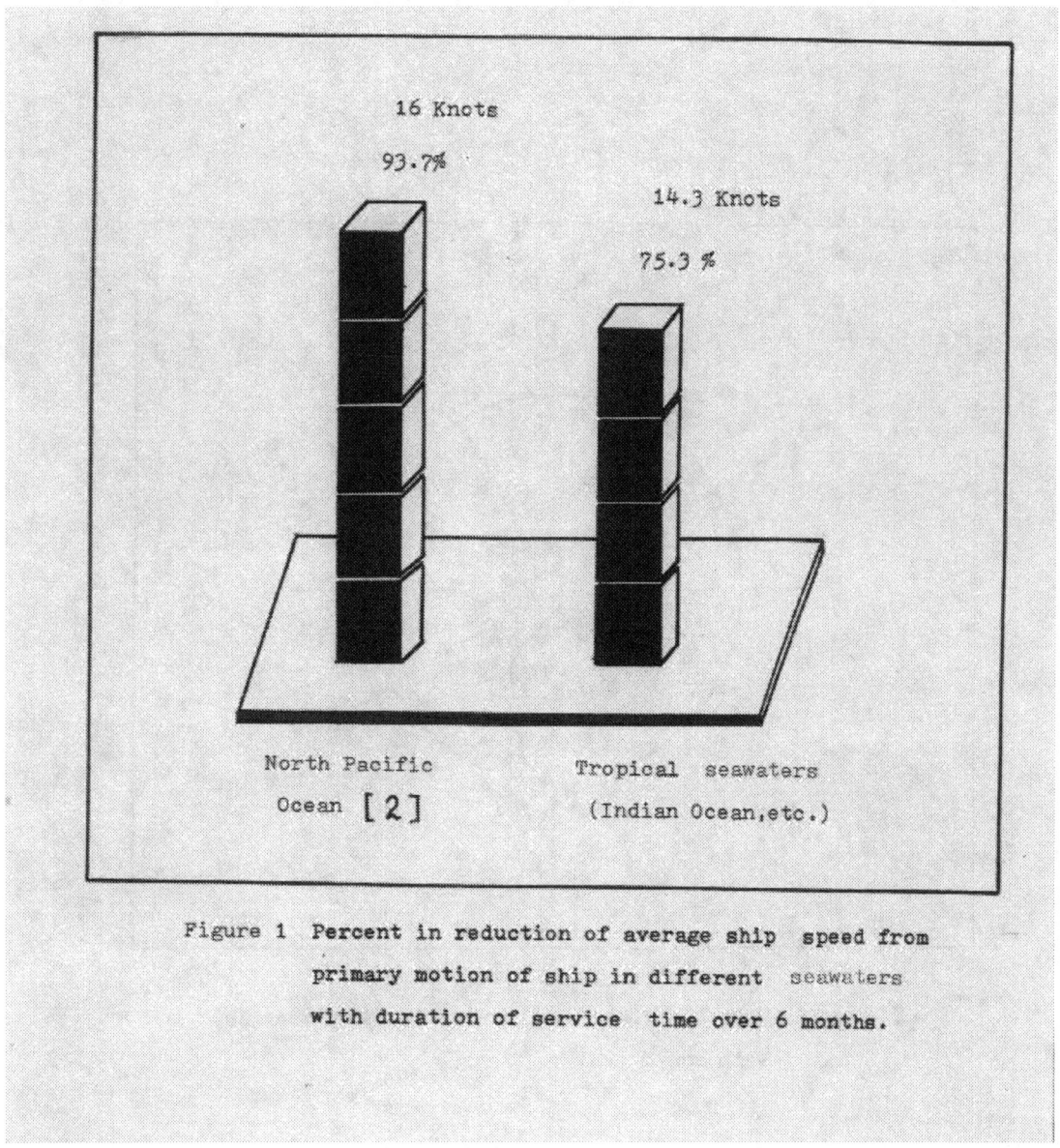

Figure 1 Percent in reduction of average ship speed from primary motion of ship in different seawaters with duration of service time over 6 months.

Figure 2 shows schematically the influence of each factor on a movable ship in the sea. However, indirectly it is possible to consider that ship speed (**Y**) is also the function of revolution per minute of the main engine (**X₁**) and has a directly proportional dependence.

With the increasing revolution of the main engine, the ship speed also increases. This functional dependence could be expressed by the equation $Y = f_1(X_1)$ (2).

Herewith, the mode of operation of the engine (its heat-strength) depends also on the revolution of the engine (X_1) and the temperature of seawaters (X_3). By the criterion of the normal operation of the engine in these conditions, we are able to use the ship as index of the temperature of exhaust gases from the engine (X_{13}).

So, it can be written by the functional equation view of $X_{13} = f_2(X_1; X_3)$ (3).

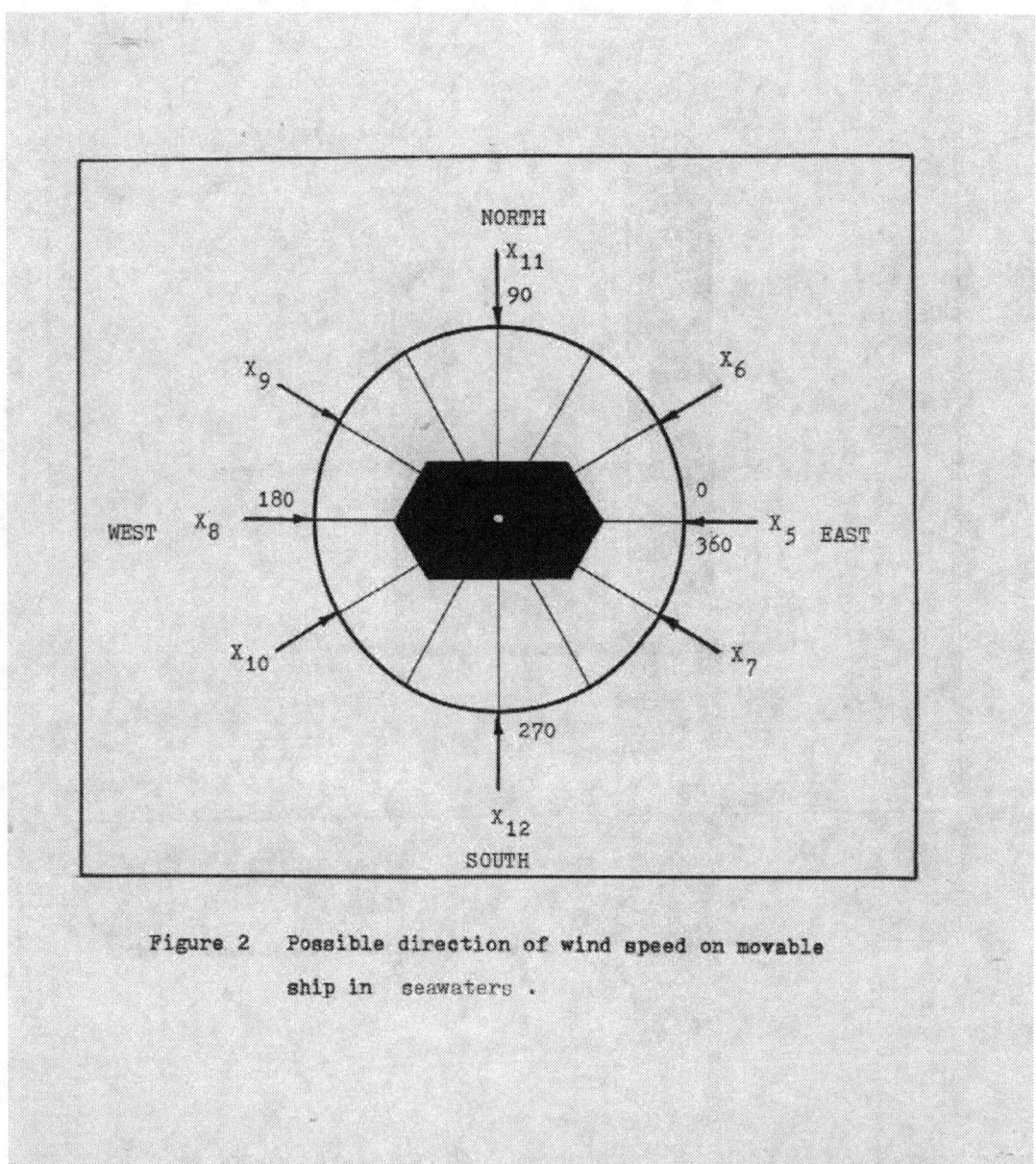

Figure 2   Possible direction of wind speed on movable

ship in   seawaters .

Analyzing the functional equations (1) ,(2) and (3) ,we see that the ship's speed (Y) is functionally joined with the above-named variables (X_1) , (X_3) and (X_{13}) ,i.e it has this view:

$$Y=f\{[\ f_1(X_1);X_2;X_3;X_4;X_5;X_6;X_7;X_8;X_9;X_{10};X_{11};X_{12};\ f_2(X_1;X_3)]\} \quad (4)$$

The functional model in total view for the ship's speed and operation of the main engine is shown in Figure 3. In an analysis of function (4) and Figure 3 ,the author makes the conclusion that a ship's speed mainly for tropical seawaters is exposed to the multiple regression analysis with the use of methods and devices of the mathematical statistics.

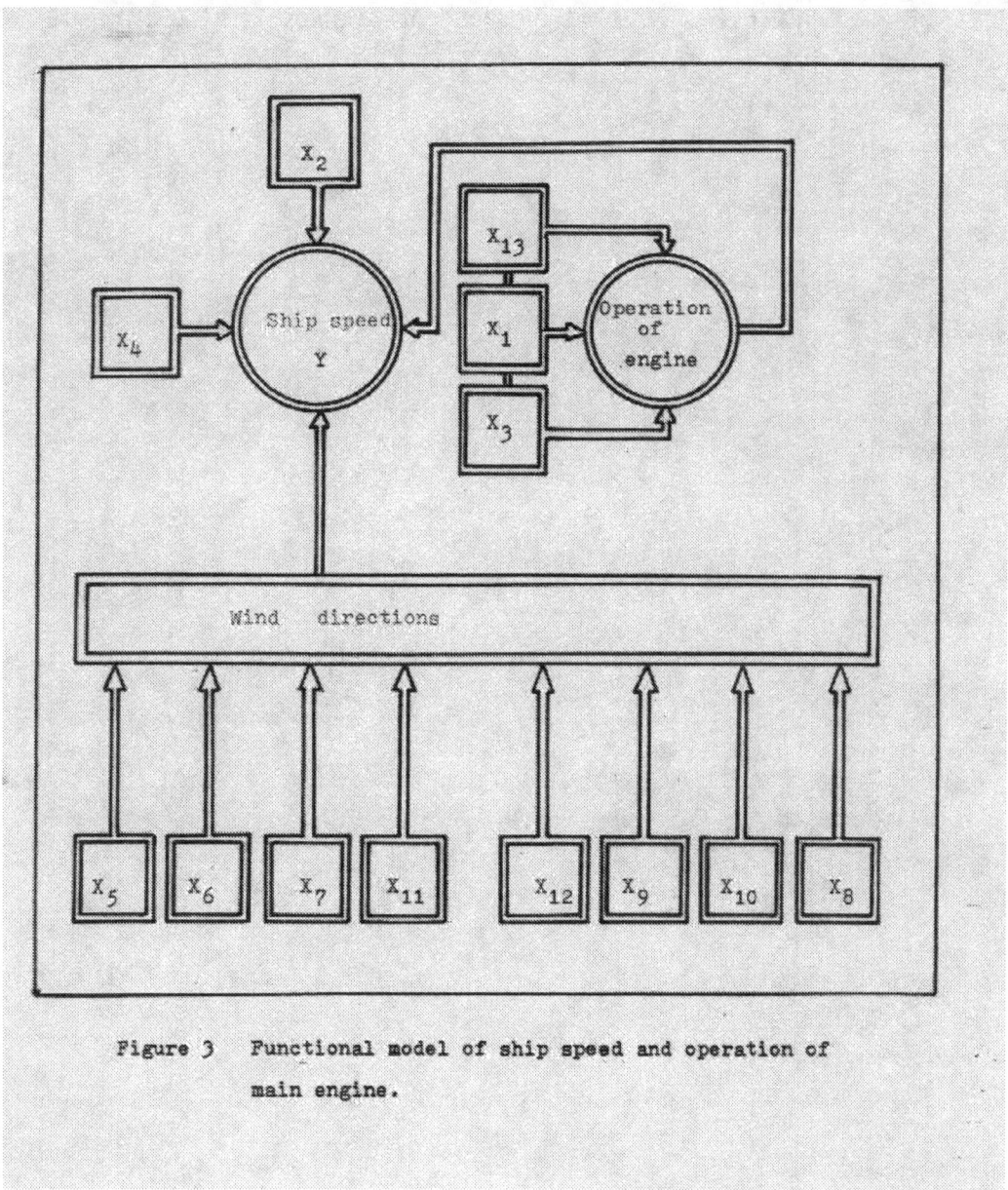

Figure 3   Functional model of ship speed and operation of
main engine.

4.3 Sharing analysis of factors having an influence on ship's speed in the tropical seawaters

It has been said above that analysis of a ship's speed is possible with the use of mathematical statistics in order to analyze and define the general factors which play an important role on changing the above-named parameters.

The author suggests the method of estimating each sharing factors n evaluation of ship's speed, and this method has the following steps:

      a. *To determine the general independent and dependent variables of experiment:*

$X_{1,1}; X_{2,1}; X_{3,i} \ldots\ldots\ldots X_{j,1}$ (independent variables)   ($i = 1,2,3 \ldots n; j = 1,2,3 .. k$)

where,

n= the number of observation (rows);
k= the number of independent variables (columns);
Y_i= dependent variable.

$b.$ *To estimate summary characteristics for each column:*

$$\sum_{i=1}^{n}X_{1,i} ; \sum_{i=1}^{n}X_{2,i}; \sum_{i=1}^{n}X_{3,i}\dots\dots\dots\dots\dots\sum_{i=1}^{n}X_{k,i} ; \text{ and } \sum_{i=1}^{n}Y_i \quad (6)$$

$c.$ *To calculate the average values of these parameters:*

$$\overline{X}= (\sum_{i=1}^{n}X_{1,i})/ n ; \overline{X}_2= (\sum_{i=1}^{n}X_{2,i})/n \dots \overline{X}_k= (\sum_{i=1}^{n} X_{k,i})/n ; \text{and } \overline{Y} = (\sum_{i=1}^{n}Y_i)/n \quad (7)$$

$d.$ *To estimate summary characteristics of independent variables for each i-value ,i.e*

$$X_{s,i} =\sum_{j=1}^{k} (X_{1,i}+ X_{2,i}+\dots X_{j,i}) \qquad X_{j,i}= \sum_{j=1}^{k} (X_{1,i} + X_{2,i}+ X_{k,i}) \quad (8)$$

$$X_{s,1}= \sum_{j=1}^{k}X_{j,1} ; X_{s,2}= \sum_{j=2}^{k}X_{j,2} ;\dots\dots.X_{s,n}= \sum_{j=1}^{k} X_{j,n}$$

$e.$ *To calculate the average of sharing dependent variable indexes of ship's speed:*

$$Y_{s,i}= Y_i/X_{s,i}= Y_i/ [\sum_{j=1}^{k} (X_{1,i}+ X_{2,i} +\dots. X_{j,i})] \quad (9)$$

$$Y_s= \sum_{i=1}^{n} (Y_{s,1}+ Y_{s,2} +\dots. Y_{s,n}) /n \quad (10)$$

$f.$ *To estimate the average percentage of sharing independent variable indexes of ship's speed:*

$$\Delta X_{1,i}= (X_{1,i}/ X_{s,i}) 100\% ; \Delta X_{2,i}=(X_{2,i}/X_{s,i}) 100\% ; \Delta X_{k,i}=(X_{k,i}/X_{s,i}\Delta) 100\%$$
(11)
and

$$\overline{\Delta X_1}=(\sum_{i=1}^{n}\Delta X_{1,i})/n ; \overline{\Delta X_2}= (\sum_{i=1}^{n}\Delta X_{2,i})/n ;\dots\dots.\overline{\Delta X_k}= (\sum_{i=1}^{n}\Delta X_{k,i})/n \quad (12)$$

59

In Table 1.1 as example indicates that the average percent of sharing indexes which are joined functionally with the ship's speed (Y'=6.08 knots is the average ship's speed accepted as 100 percent).

Table 1.1 Calculation of sharing percent indexes are joined functionally with the ship's speed

ΔX_1	ΔX_2	ΔX_3	ΔX_4	ΔX_5	ΔX_6	ΔX_7	ΔX_8	ΔX_9	ΔX_{10}	ΔX_{11}	ΔX_{12}	ΔX_{13}
6.02	42.09	13.93	0.16	8.11	0.23	2.23	0.74	0.44	0.71	0.40	1.42	23.47

Figure 4 shows a characteristic of each parameter functionally influencing on the ship's speed.

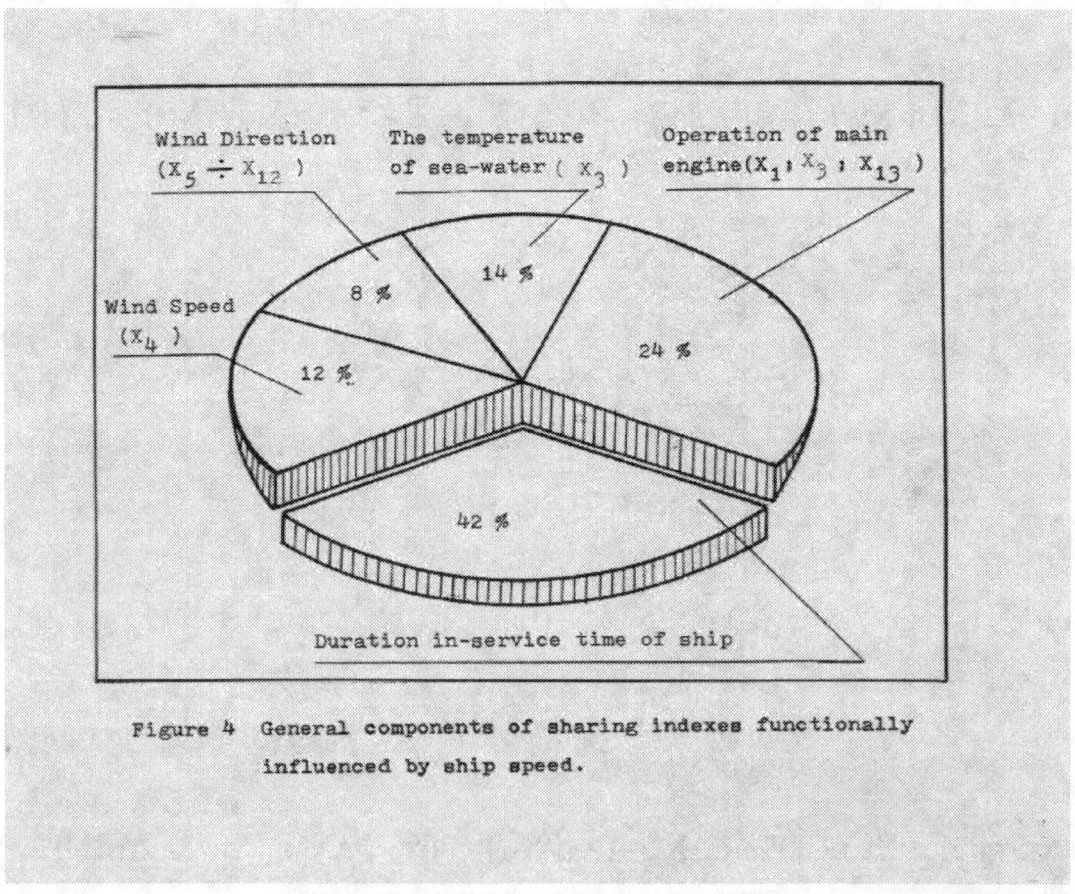

Figure 4 General components of sharing indexes functionally influenced by ship speed.

Analysis of Figure 4 indicates the one fact that the most influences on the reduction of a ship's speed is the duration of in-service time of a ship (X₂);the sharing percent index is joined functionally with the ship's speed and is equal to **ΔX_2=42.09 percent,** and particularly this is the important for the ship ,sailing in the tropical seawaters than in the cold seawaters.

Not the least of the factors is the temperature of exhaust gases from the engine (X_{13}) and the temperature of seawater inputting in the engine as an outer cooling system (X_3) ,i.e the operation process of the main engine because the sharing indexes functionally influencing on the ship's speed accordingly are equal to $\Delta X_3 = 13.98$ percent and $\Delta X_{13} = 23.47$ percent.

4.4 Statistical analysis of general parameters of motion ship and operation of the main engine

Characteristic peculiarities of the operation of the main engine and a ship in the conditions of sailing in the tropics becomes more obvious .

First of all, because with the duration of in-service time of a ship, the revolution of the engine and the ship's speed decrease accordingly as this is shown in Figure 5.

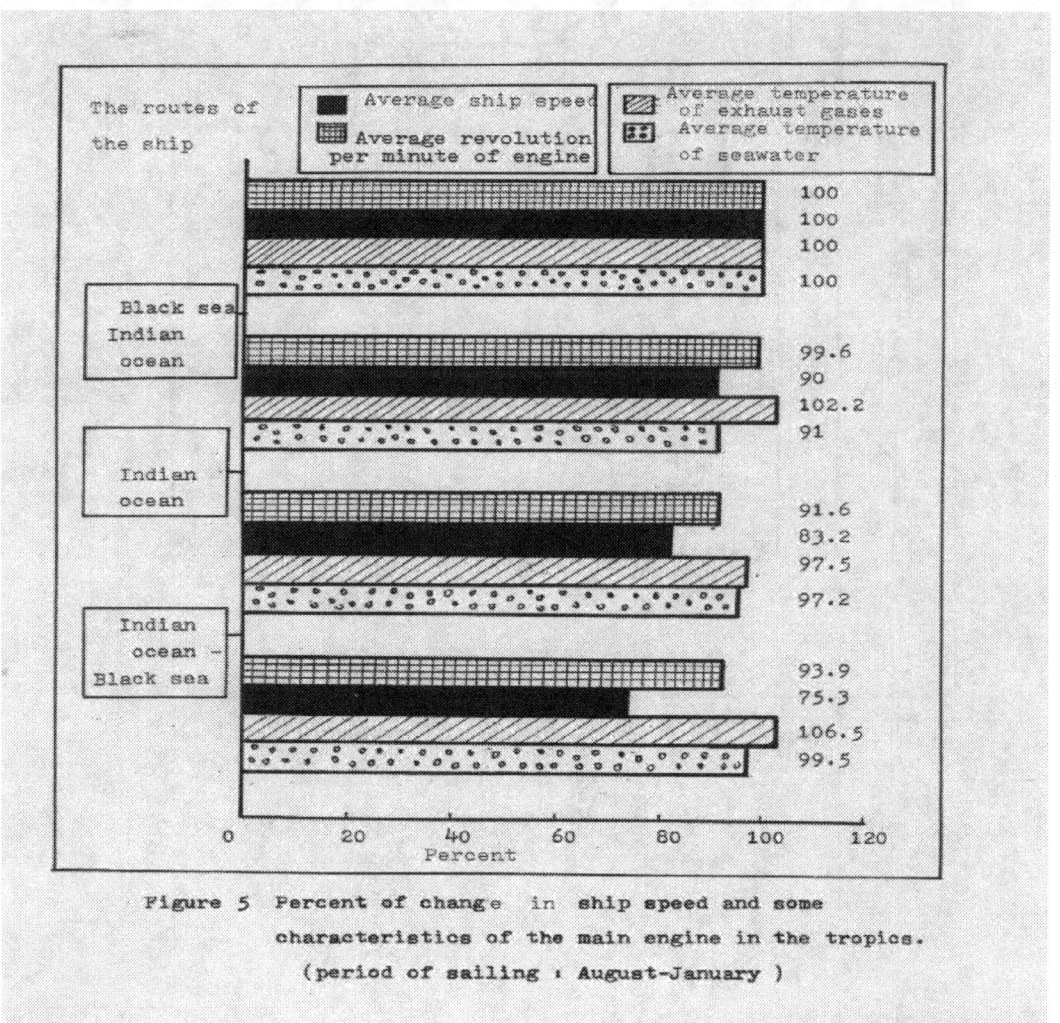

Figure 5 Percent of change in ship speed and some characteristics of the main engine in the tropics. (period of sailing : August-January)

61

In analyzing of Figure 5 we see for the period in which ship sailing in the tropical seawaters for more than six months, the average revolution of the main engine decreases more than **6 percent** in comparison with its starting value of **n=110 revolutions per min.**

Otherwise leaving the revolution of the main engine constant for the duration of the whole trip will be raised the temperature of the exhaust gases from the engine; i.e the heat-stress of the engine will increase and the service life of the engine will decrease as a result of this action .

Therefore, in practical conditions for the ship sailing in the tropical seawaters , one should periodically to correct the revolution of the main engine until the value of the present temperature of exhaust gases will be normal in the working processes of the main engine.

The process of correction the revolution of the main engine is shown in Figure 6 as a process of sailing a ship in the tropics for a period more than six months.

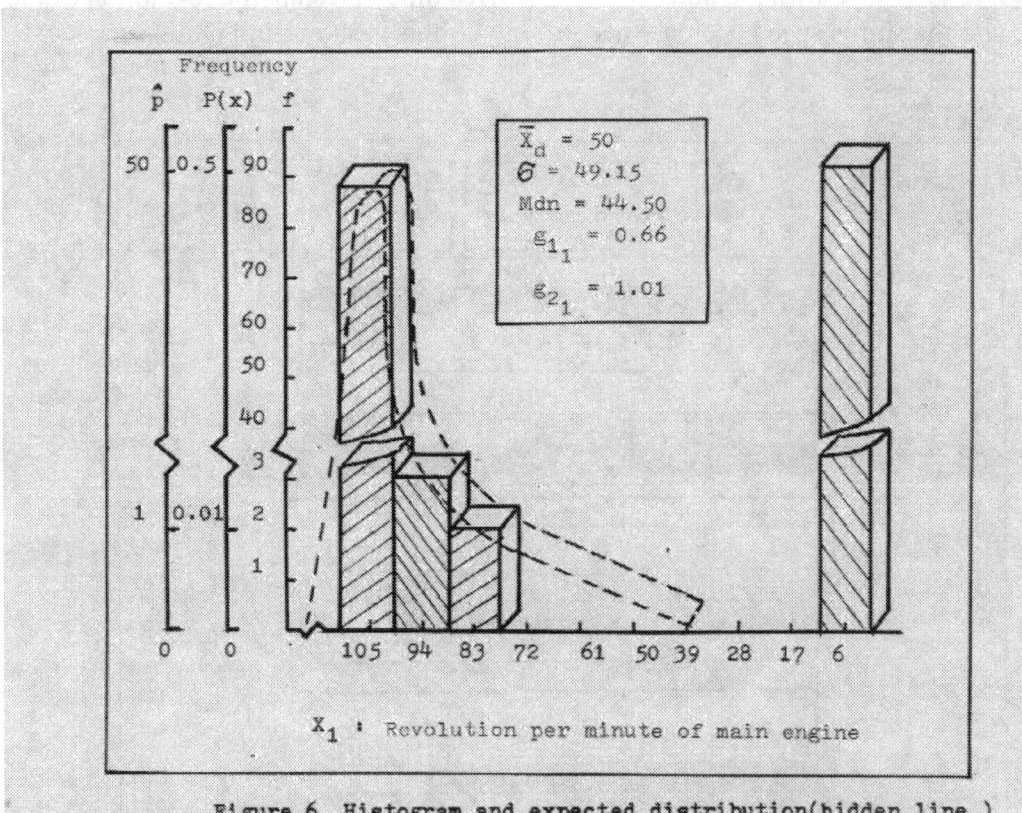

Figure 6 Histogram and expected distribution(hidden line)
of revolution of the main engine (X_1) with
relative P(x),percentage ($\hat{p}$) and frequency (f)
axes.

Analyzing Figure 4 ,we see that the duration of in-service time of the ship (X_2)is the most important factor of decreasing of ship speed.

A characteristic peculiarity of the histogram and expected distribution, as this is shown in Figure 7, of this parameter (X_2) shows that the frequency is equal to **f=1**,relative frequency is equal to **P(x)=0.0055,**and percentage of frequency is equal to **p'=0.55 percent.**

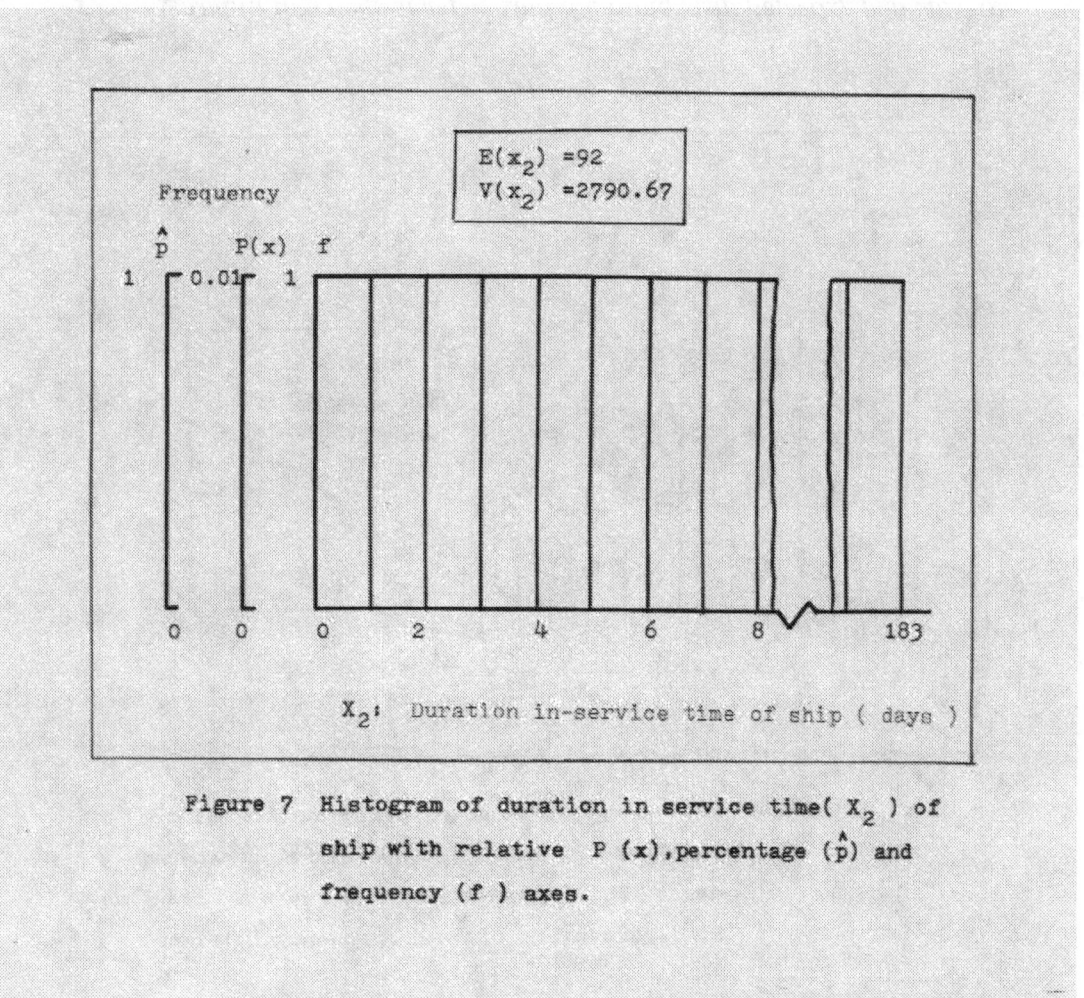

Figure 7   Histogram of duration in service time(X_2) of ship with relative P (x),percentage ($\hat{p}$) and frequency (f) axes.

This will characterize the histogram distribution as discrete (rectangular) symmetric distribution with the following characteristics:

$$Mean:\ E(X_2) = \sum_{i=1}^{n} X_2\, P(x) = \sum_{i=1}^{n} X_2(1/n) = (1/n) \sum_{i=1}^{n} X_2 \qquad (13)$$

where,
n= the number of observations.
Variance:

$$V(X_2) = \sum_{i=1} [X_2 - E(X_2)] \; P(x) = (1/n) \sum_{i=1} [X_2 - E(X_2)] \qquad (14)$$

It is necessary to admit that the equally important characteristic in the evaluation of the ship's motion is the regime of operation of the main engine, namely ,as the revolution of engine (X₁)examined above in Figure 6 and the temperature of exhaust gases (X₁₃)from the main engine.

The histogram and expected distribution of parameter is shown in Figure 8.

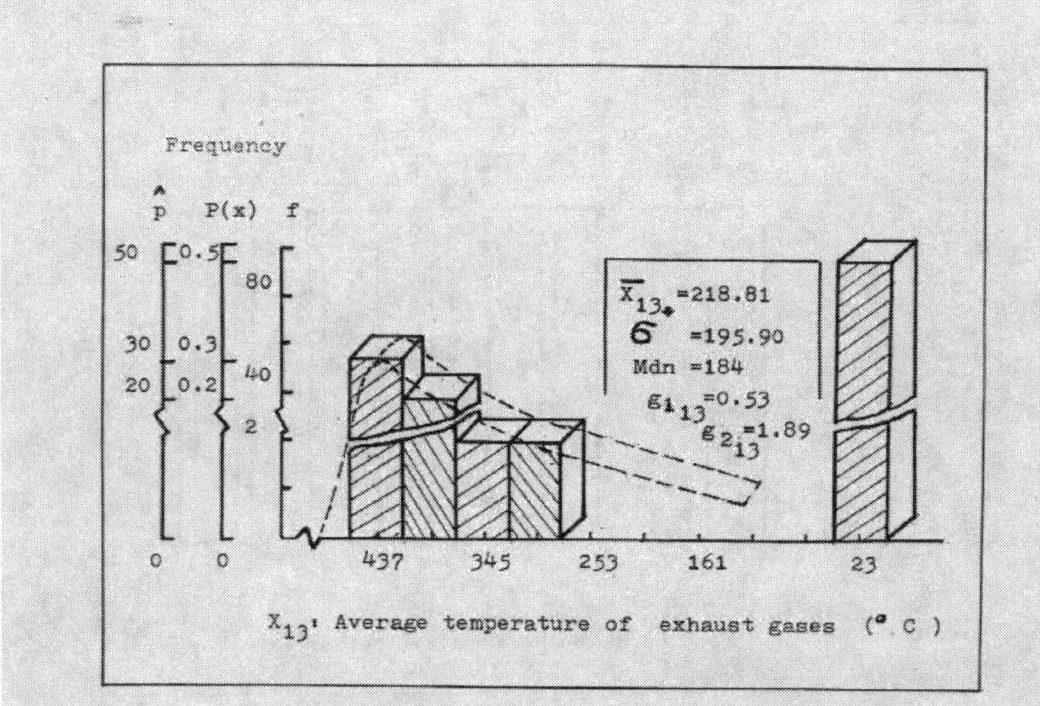

Figure 8 Histogram and expected distribution (hidden line) of the average temperature of exhaust gases (X_{13}) from the main engine with relative $P(x)$,percentage $(\hat{p})$ and frequency (f) axes.

Analysis of the histogram and expected distributions as shown in Figures 6 and 8 indicates the similarity of their distributions. In both cases, the character of these distributions submits to the rules of positive skewed right distribution with the parameters:

$$g_1 = [3 \, (\overline{X_{j*}} - Mdn)] / \sigma \; (15)$$

where,
g₁=skewness;

$\overline{}$

X_{j*}= the arithmetical mean from group data (j=1,2,3.......n) ,equal

$$\overline{X_{j*}} = \overline{X_d} + (\textstyle\sum f\, d\,)/n \qquad (16)$$

$\overline{X_d}$= the selected mid-value of any class;

d= deviation from assumed mean;

σ=standard deviation, grouped data, equal $\qquad \sigma=[\,(\sum f\, X_i\,)/n\,]^{2\ 0.50} \qquad (17)$

where,

X_i= deviation of mid-values I-class from $\overline{X_{j*}}$,equal $\quad X_i = (\overline{X_d} - \overline{X_{j*}}) \qquad (18)$

Analysis of Figure 6 and 8 shows that their distributions have peaked more than normal and for this reason such distribution will be referred to as the leptokurtic distribution with

estimation parameter such as kurtosis:

$$g_2 = (\textstyle\sum f\, X^4\,)\,/\,(n\sigma^4) \qquad (19)$$

Conformably to Figure 6 we have the following:

X_d=50 ; $\sum fd$=979 ; n=183; X_{1*}=55.35; $\sum f\, x^2 = 4.42 \cdot 10^5$;$\sum f\, x^4 = 1.08 \cdot 10^9$;$\sigma = 49.15$; Mdn=44.50; $g_{1,1}$=0.66;$g_{2,1}$=1.01.

Conformably to Figure 8 we have the following:

X_d=207; $\sum f\, d$=2162; n=183;X_{13*}=218.81 ; $\sum f\, x^2 = 7.02 \cdot 10^6$; $\sum f\, x^4 = 2.797 \cdot 10^{11}$; σ=195.90 ; Mdn= 184; $g_{1,13}$=0.53; $g_{2,13}$=1.89.

*From Figure 6 we see that the greater percentage of frequency of distribution (**p' =49.70 percent**) falls on the period of mooring the ship in the harbor with tropical seawaters (the revolution of the main engine in this period was equal to zero).*

As a result of these actions, the body of the ship was encrusted significantly with marine growth. The same picture has been placed in Figure 8,where it shows the intercommunication of temperatures of the exhaust gases from the revolution of the main engine. With decreasing revolution (X_1) of the engine ,the temperature of exhaust gases (X_{13}) increase accordingly. As was shown above ,the temperature of seawater indirectly influences the changing of ship speed and operation of the main engines, therefore ,the studying of this parameter deserves more attention.

Figure 9 shows the histogram and view of the expected distribution of this parameter (X_3).

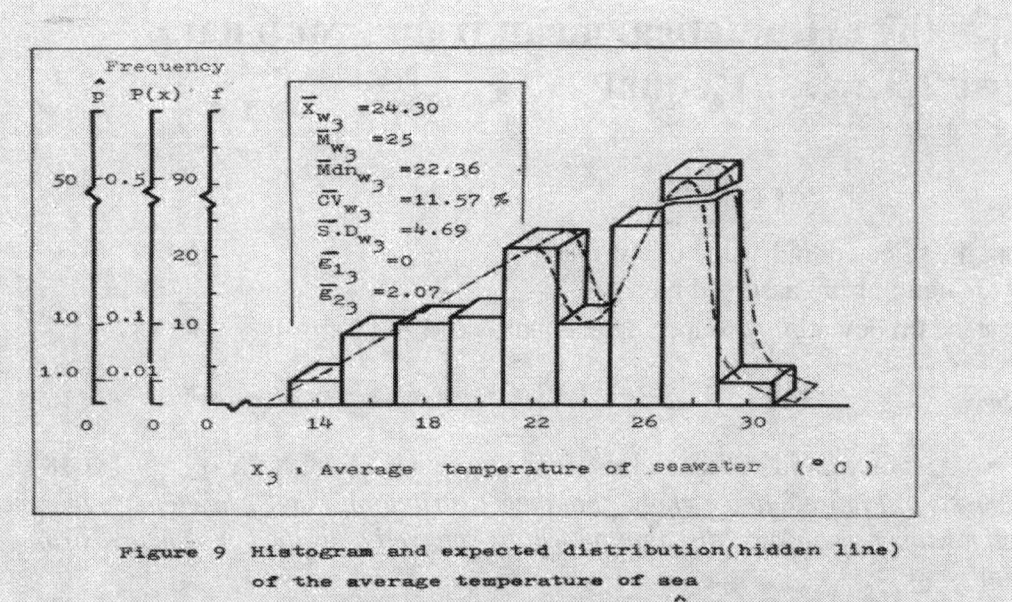

Figure 9   Histogram and expected distribution(hidden line)
of the average temperature of sea
with relative P(x),percentage ($\hat{p}$) and frequency
(f) axes.

Analysis of Figure 9 shows that the shape of the frequency distribution is bimodal consisting of two peaks at this complex distribution which has generalized the characteristics:

The average value	$X_{w,3}$=24.30;
Mode	$M_{w,3}$=25;
Median	$Mdn_{w,3}$=22.36;
Coefficient of variation	$CV_{w,3}$=11.57 percent;
Standard deviation	$S.D_{w,3}$=4.69;
Skewness	$g_{1,3}$=0;
Kurtosis	$g_{2,3}$=2.07

As the value of skewness is equal to $g_{1,3}=0$ analysis of Figure 9 indicates that the symmetric distribution and the value of kurtosis is equal to $g_{2,3}=2.07$ and indicates on the considerable flat of the curve of polygon distribution on the side of the right part of this distribution.

So, the author of this paper investigates, on the basis of statistical analysis, three general characteristics which influence on the operational of the main engine:

   a.   **The revolution per minute of the engine;**
   b.   **The temperature of exhaust gases from the engine;**
   c.   **The temperature of seawaters in the tropics.**

All these parameters were described and shown in Figure 6 and Figure 8 which have abnormal distribution with the positive right-side skewed and bimodal distribution shown and described in Figure 9 with two peaks.

Variation of ship speed for the period of duration of in-service time of the ship is shown in Figure 10.

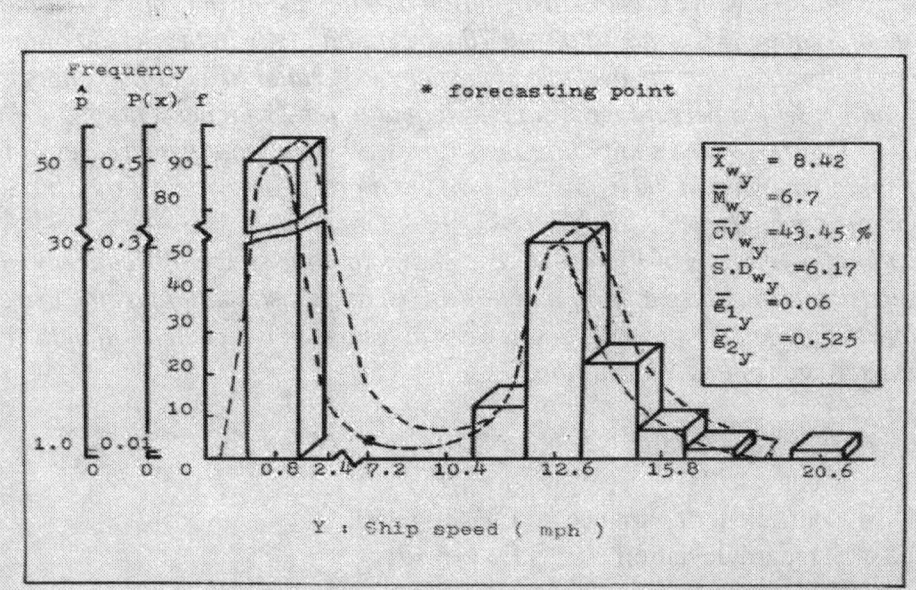

Figure 10 Histogram and expected distribution (hidden line)
of the average ship speed (Y) with relative P(x).
percentage (p̂) and frequency (f) axes.

*Analysis of Figure 10 with expected distribution and the histogram shows that the percentage of frequency is equal to **p'=49.70** percent and falls in the period of the ship standing in the harbor, and the average ship speed was equal to **Ys=10.95** knots for the period of motion with a percentage frequency equal to **p'=27.80** percent .*

Analysis of Figure 10 also shows that the ship speed significantly decreases in the period of sailing the ship in the tropical seawaters, particularly after of the ship was standing in the harbor.

And besides, the analysis of Figure 10 also shows that the shape of frequency in data is distributed as abnormal distribution and has a U-shaped distribution, as the extreme values have the highest frequencies and fewer values are in the center which have the total characteristics of this distribution:

The average value	$X_{w,y}$=8.42;
Mode	$M_{w,y}$=6.70;
Coefficient of variance	$CV_{w,y}$=43.50;
Standard deviation	$S.D_{w,y}$=6.17;
Skewness	$g_{1,y}$ = 0.06;
Kurtosis	$g_{2,y}$=0.525.

Therefore, form Figure 10 we see that the ship speed in during of sailing in the tropics has an abnormal distribution and the frequency polygon has the complex U-shaped distribution.

Besides the other factors, the most important role in reduction of ship speed is the wind speed and its directions on the ship in during of motion on the sea. This fact is shown in the histogram and expected distribution of these indexes as in Figure 11 and Figure 12.

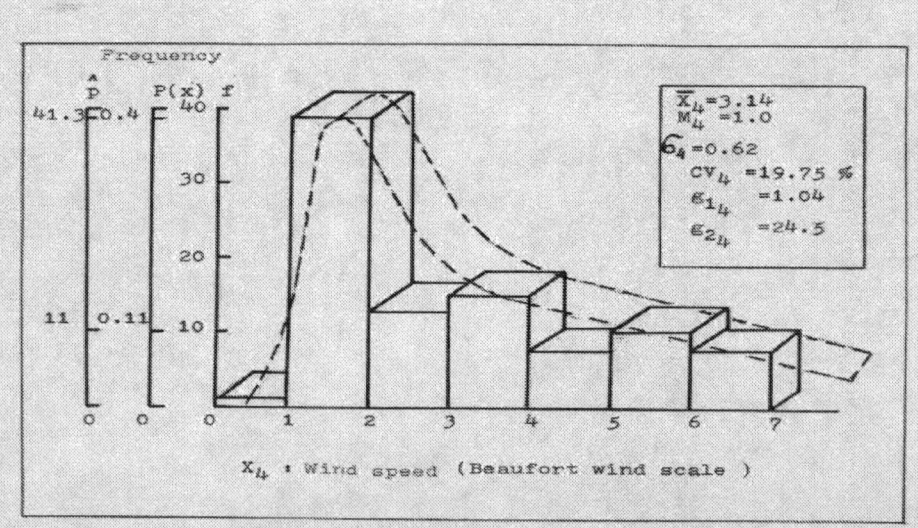

Figure 11   Histogram and expected distribution of wind
speed (X_4) with relative P(x),percentage ($\hat{p}$)and
frequency (f) axes.

Analysis of the expected distribution, as shown in Figure 11,indicates that its curves has an abnormal distribution with the positive right skewness in the following parameters:

The average value	$\bar{X}_4 = 3.14$;
Mode	$M_4 = 1.0$;
Standard deviation	$\sigma_4 = 0.62$;
Coefficient of variation	$CV_4 = 19.75$ percent;
Skewness	$g_{1,4} = 1.04$;
Kurtosis	$g_{2,4} = 24.50$

As the value of kurtosis is equal to $g_{2,4}=24.50>3.0$ this is indicated on a most considerable peaked curve of this distribution and the value of skewness is equal to $g_{1,4}=1.04$ and this characterizes a slight positive skew.

The histogram and expected distribution of the data of wind speed are shown in Figure 12 which holds a significant influence on reduction of ship speed and operation of the main engine.

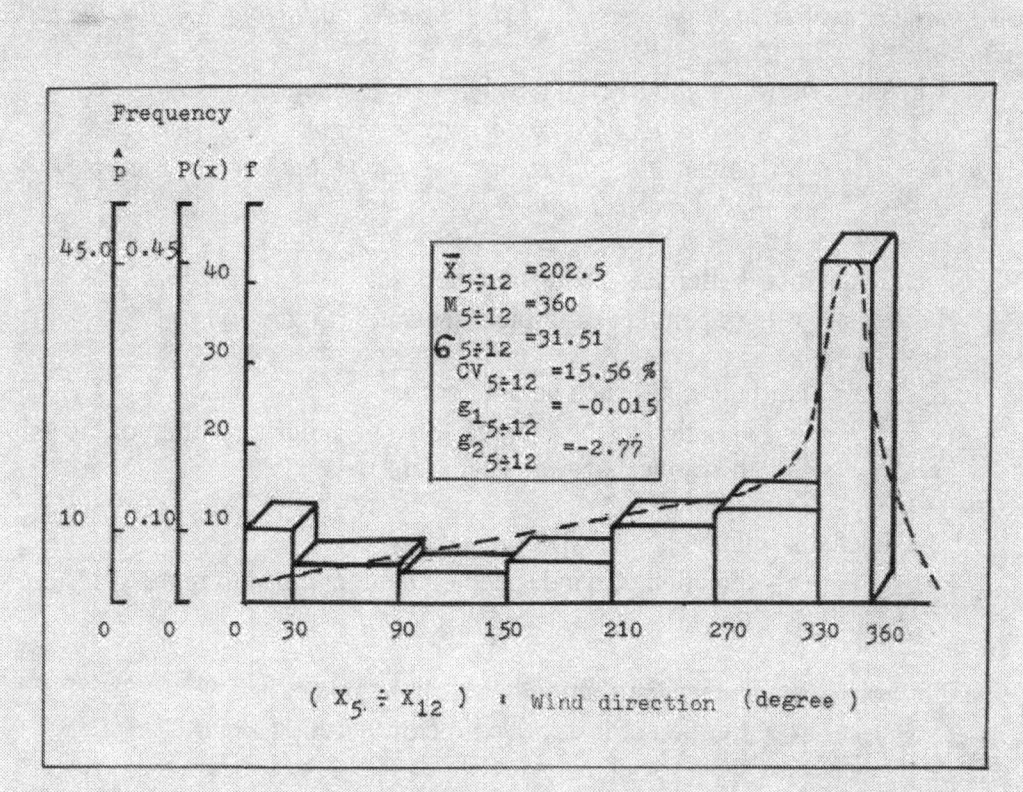

Figure 12   Histogram and expected distribution (hidden line) of direction of wind ($X_5 \div X_{12}$) with relative P(x),percentage (p) and frequency (f) axes.

Anatoly Rozenblat

As we see from Figure 12, the extreme values reach the right side of this distribution ,and all this shows that the frequency of the polygon relates to a J-shaped distribution in the following parameters:

The average value $\qquad$ $X_{5÷12}$ =202.50;
Mode $\qquad$ $M_{5÷12}$=360;

Standard deviation $\qquad$ $\sigma_{5÷12}$=31.51;
Coefficient of variation $CV_{5÷12}$=15.56 percent;
Skewness $\qquad$ $g_{1,5÷12}$= 0.015;
Kurtosis $\qquad$ $g_{2,5÷12}$=-2.77.

The negative value of skewness ($g_{1,5÷12} = 0.015$) indicates a slight negative skew and the value of kurtosis is equal to ($g_{2,5÷12}= -2.77$) and this indicates a slightly platykurtic distribution.
   Statistical analysis of ship speed and some characteristics of the main engine are shown in the above-named Figures 6,7,8,9,10 ,11 and 12 allowing to make the following conclusions:
   a. Positive skewed right distributions have such functional parameters as :
- *revolution of the main engine (Figure 6);*
- *temperature of exhaust gases from the engine (Figure 8);*
- *wind speed (Figure 11).*

   b. Bimodal distribution has such parameters as;
- temperature of tropical seawater (Figure 9).

   c.U-shaped distribution has such parameters as:
- variation of the average ship speed during sailing of the ship in the tropical seawaters (Figure 10);

   d. J-shaped distribution has such parameters as :
- direction of wind speed on the moving ship in the sea (Figure 12);

   e. Discrete (rectangular) distribution has such parameter as :
- duration of in-service time of the ship (Figure 7).

Conclusions and recommendations

a. *The main objective of this paper was focused on the statistical analysis of ship speed reduction in conditions of sailing ship in the tropical seawaters . General attention*

70

was devoted to the functional analysis of parameters influencing on the ship speed and also to the operation of the main engine;

b. *The results of the paper are shown in view of histograms and expected distributions for each variable parameter, indicating the most important characteristics of these distributions in a generalized view;*

c. *The statistical analysis indicates that the ship speed after of six months of sailing in the tropical seawaters has a **30 percent reduction** from the starting ship speed. This reduction of ship speed was reached by means of duration of the in-service time of the ship (more than six months of sailing in the tropical seawaters). An indirect index of this event is in the reduction of ship speed as a result of marine fouling on the body of this ship;*

d. *A similar picture focused on the main engine indicates that the revolution of the engine decreases to **6 percent** at the same conditions;*

The author in this paper would like to give some recommendations regarding of improving the operational of the main engine and also of increasing of the ship speed in the process of sailing ship in the tropical seawaters in view of :

(1) After of six month navigation process in the tropical seawaters each merchant ship necessary to arrange in the dock for removing of marine fouling from the its body;

(2) To design the more effective outer cooling system for any main engine which uses in the tropical seawaters with the objective of decreasing of its heat-stress and normal working conditions;

(3) As a result ,the rising of temperature of exhaust gases in the of operational process of main engine is as indirectly index of appearing of the marine fouling on the body of the ship and these peculiarities should to include in service duties for the operational people that in time to reduce the engine speed and guarantee the good heat-stress of this main engine.

References

[1] V.V Zvonkov," Marine Tow Tractor Calculations ",(Moscow: River Transport Publisher,1956): 51-71.

[2] P. Lacey and R. Edwards," ARCO Tanker Slamming Study", Marine Technology", *(July,1993):,135-147.*

[3] Y.Washio, M.Miyoshi, K.Takekuma, K.Vamada, and Kobayashi, " Recent Research and Development in the Design of an Oceanographic Research Vessel", Marine Technology ,(January ,1994): 1-19.
[4] B.W. Parker , " Marine Transportation in Alaska's Bering Sea and Artic Ocean Areas", Proceedings of the First International Conference on Port and Ocean Engineering under Artic Conditions, (*Trondheim, Norway: Technical University of Norway ,1972),1:790-802.*

[5] G. Vossers, Resistance, Propulsion and Steering of Ships,(*Antwerp-Cologne: The Technical Publishing Company H. Stam N.Vhaarlem,1962) 11 C:83;*

[6] R.M Crosby ," Ocean Instrumentation: The Better Use of Materials Technology",1970 IEEE International Conference on Engineering in the Ocean Environment (September ,1970): 147-149;

[7] R. Balasurbramanyan, N.Unnikrishnan Nair, and A.G.Gopalakrisna Pellai," The problem of Marine Fouling in the Coastal Waters of India and its Economic Imlications with Special Reference to Fishing Fleet Management", Proceedings: Third International Congress on Marine Corrosion and Fouling,(Evanston ,Illinois: Northwestern University Press,1972): 898.

[8] J. Sandison,D. Woolaver, M.Dipper, and M.Rice, "Sea Trials of the SWATH Ship USNS Victorious (T-AGOS 19), " Marine Technology ", (October ,1994),31:245-257.**

Bibliography

Croxton, Frederick E.and Cowden, Dudley J., Applied General Statistics, *New York: Prentic-Hall, Inc. 1939*

*Havilcek,L.L and Crain,R.D .,*Practical Statistics for the Physical Sciences, *Washington, D.C.,: American Chemical Society ,1988.*

Heywood ,John B., Internal Combustion Engine Fundamentals ,*New York : McGraw-Hill Book Company,1988.*

Kreyszig,Erwin, Introductory Mathematical Statistics, *New York: John Wiley&Sons,1970*

Meyer, Stuart L., Data Analysis for Scientists and Engineers, *New York: John Wiley& Sons,1975.*

Pfaffenberer, Roger C. and Patterson, James H., Statistical Methods ,*Homewood, Illinois: Richard D.Irwin,Inc.,1977.*

Schmid, Calvin F. Statistical Graphics ,*New York: John Wiley&Sons,1983.*

CHAPTER 5

Applied Multiple Regression Analysis of Ship speed in the tropics

5.1 The general multiple regression model for the ship of speed

Multiple regression analysis is one of the most widely used of all statistical methods and it also highly useful in experimental and research situation when necessary to decrease the big capital investments and find primary the general ways for further investigations.

These conditions are put in many research works in question of increasing of ship speed. Many researchers in his reviews pay the big attention to the facts only in the construction of new ships and its modernization which will improve the technico-economical characteristics of ship advantageously the ship speed as the main index of its profitable.

And besides these papers describe advantageously the regression models having the functional analysis view of Y=φ(X) with one independent variable, i.e view of

$$Y_i = b_0 + b_1 X_{i,1} + e_i \quad (I=1,2\ldots n) \qquad (1)$$

However, as show the further investigations and analysis the present observation data, the ship speed is the complex parameter which depends from many independent predictor variables and submits to the multiple regression analysis.

In present paper is given the multiple regression model for the ship speed into account of the different factors which act on this main index in period of its operation in the tropics and wave sea.

The purpose of this research is the investigation of ship speed in the tropics and wave sea in dependence from some independent variables such as:

- X_1=revolution per minute of the main engine;
- X_2 =duration in-service time of ship, days;
- X_3= temperature of seawater of outer cooling system of the main engine,°C;
- X_4=wind speed (Beaufort wind scale);
- X_5= wind direction on the east (E) ,degree;
- X_6=wind direction on north –east (NE) ,degree;
- X_7=wind direction south-east(SE),degree;
- X_8=wind direction on west (W) ,degree;
- X_9=wind direction in north-west (NW),degree;
- X_{10}=wind direction on south-west (SW) ,degree;
- X_{11}=wind direction on north (N) ,degree;

- X_{12}=wind direction on south(S),degree;
- X_{13}=temperature of exhaust gases from engine,°C;

The dependent variable Y is the ship speed (in miles per hour).

Figure 1.1 shows the abstract scheme of forces which act on a ship in the motion.

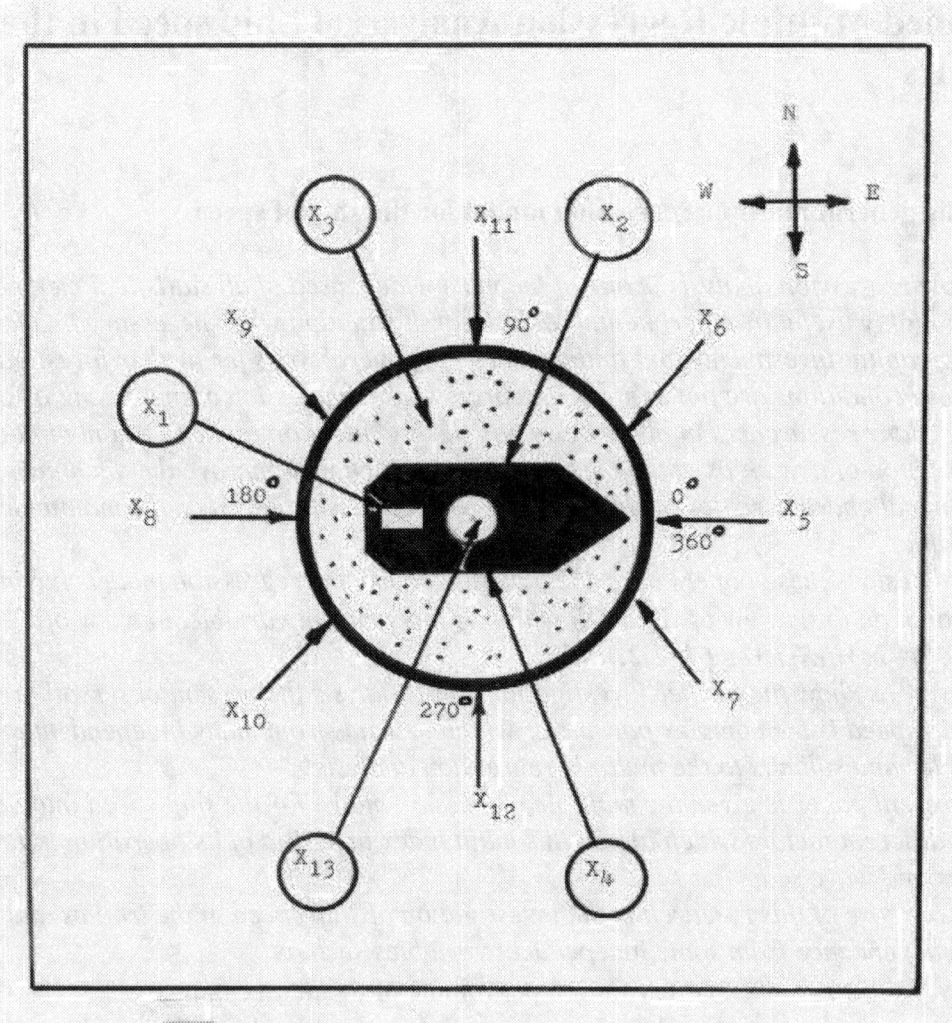

Figure 1.1 Abstract scheme of force actions on a ship
in the motion

Multiple regression model for the ship speed can be expressed as

$$Y = b_0 + b_1 X_1 + b_2 X_2 + \ldots\ldots\ldots b_{13} X_{13} \qquad (2)$$

where,

$b_0, b_1, b_2 \ldots \ldots b_{13}$ *the coefficients of estimation and* $X_1, X_2 \ldots \ldots \ldots X_{13}$ *the independent variables.*

To put the matter concretely in our present problem, we must write the estimating equation in view of:

$$\hat{Y} = b_0 + b_1 X_1 + b_2 X_2 + b_3 X_3 + b_4 X_4 + b_5 X_5 + b_6 X_6 + b_7 X_7 + b_8 X_8 + b_9 X_9 + b_{10} X_{10} + b_{11} X_{11} + b_{12} X_{12} + b_{13} X_{13}$$

$$(3)$$

which requires simultaneous solution of the 14 normal equations.

1. $\sum Y = n b_0 + b_1 \sum X_1 + b_2 \sum X_2 + b_3 \sum X_3 + b_4 \sum X_4 + b_5 \sum X_5 + b_6 \sum X_6 +$
$+ b_7 \sum X_7 + b_8 \sum X_8 + b_9 \sum X_9 + b_{10} \sum X_{10} + b_{11} \sum X_{11} + b_{12} \sum X_{12} + b_{13} \sum X_{13}$ **(4)**

2. $\sum X_1 Y = b_0 \sum X_1 + b_1 \sum X_1^2 + b_2 \sum X_1 X_2 + b_3 \sum X_1 X_3 + b_4 \sum X_1 X_4 + b_5 \sum X_1 X_5 + b_6 \sum X_1 X_6 +$
$+ b_7 \sum X_1 X_7 + b_8 \sum X_1 X_8 + b_9 \sum X_1 X_9 + b_{10} \sum X_1 X_{10} + b_{11} \sum X_1 X_{11} + b_{12} \sum X_1 X_{12} + b_{13} \sum X_1 X_{13}$

$$(5)$$

3. $\sum X_2 Y = b_0 \sum X_2 + b_1 \sum X_2 X_1 + b_2 \sum X_2^2$
$+ b_3 \sum X_2 X_3 + b_4 \sum X_2 X_4 + b_5 \sum X_2 X_5 + b_6 \sum X_2 X_6 +$
$+ b_7 \sum X_2 X_7 + b_8 \sum X_2 X_8 + b_9 \sum X_2 X_9 + b_{10} \sum X_2 X_{10} + b_{11} \sum X_2 X_{11} + b_{12} \sum X_2 X_{12} + b_{13} \sum X_2 X_{13}$

$$(6)$$

4. $\sum X_3 Y = b_0 \sum X_3 + b_1 \sum X_3 X_1 + b_2 \sum X_3 X_2 + b_3 \sum X_3^2 + b_4 \sum X_3 X_4 + b_5 \sum X_3 X_5 + b_6 \sum X_3 X_6 +$
$+ b_7 \sum X_3 X_7 + b_8 \sum X_3 X_8 + b_9 \sum X_3 X_9 + b_{10} \sum X_3 X_{10} + b_{11} \sum X_3 X_{11} + b_{12} \sum X_3 X_{12} + b_{13} \sum X_3 X_{13}$

$$(7)$$

5. $\sum X_4 Y = b_0 \sum X_4 + b_1 \sum X_4 X_1 + b_2 \sum X_4 X_2 + b_3 \sum X_4 X_3 + b_4 \sum X_4^2 + b_5 \sum X_4 X_5 + b_6 \sum X_4 X_6 +$
$+ b_7 \sum X_4 X_7 + b_8 \sum X_4 X_8 + b_9 \sum X_4 X_9 + b_{10} \sum X_4 X_{10} + b_{11} \sum X_4 X_{11} + b_{12} \sum X_4 X_{12} + b_{13} \sum X_4 X_{13}$

$$(8)$$

6. $\sum X_5 Y = b_0 \sum X_5 + b_1 \sum X_5 X_1 + b_2 \sum X_5 X_2 + b_3 \sum X_5 X_3 + b_4 \sum X_5 X_4 + b_5 \sum X_5^2 + b_6 \sum X_5 X_6 +$
$+ b_7 \sum X_5 X_7 + b_8 \sum X_5 X_8 + b_9 \sum X_5 X_9 + b_{10} \sum X_5 X_{10} + b_{11} \sum X_5 X_{11} + b_{12} \sum X_5 X_{12} + b_{13} \sum X_5 X_{13}$

$$(9)$$

7. $\sum X_6 Y = b_0 \sum X_6 + b_1 \sum X_6 X_1 + b_2 \sum X_6 X_2 + b_3 \sum X_6 X_3 + b_4 \sum X_6 X_4 + b_5 \sum X_6 X_5 + b_6 \sum X_6^2 +$

$$+b_7\sum X_6 X_7+b_8\sum X_6 X_8+b_9\sum X_6 X_9+b_{10}\sum X_6 X_{10}+b_{11}\sum X_6 X_{11}+b_{12}\sum X_6 X_{12}+b_{13}\sum X_6 X_{13}$$

$$(10)$$

8. $\sum X_7 Y=b_0\sum X_7+b_1\sum X_7 X_1+b_2\sum X_7 X_2+b_3\sum X_7 X_3+b_4\sum X_7 X_4+b_5\sum X_7 X_5+b_6\sum X_7 X_6$
$+$

$+$

$$b_7\sum X_7+b_8\sum X_7 X_8+b_9\sum X_7 X_9+b_{10}\sum X_7 X_{10}+b_{11}\sum X_7 X_{11}+b_{12}\sum X_7 X_{12}+b_{13}\sum X_7 X_{13}$$

$$(11)$$

9. $\sum X_8 Y=b_0\sum X_8+b_1\sum X_8 X_1+b_2\sum X_8 X_2+b_3\sum X_8 X_3+b_4\sum X_8 X_4+b_5\sum X_8 X_5+b_6\sum X_8 X_6$
$+$

$$+\,b_7\sum X_8 X_7+b_8\sum X_8^2$$
$$+b_9\sum X_8 X_9+b_{10}\sum X_8 X_{10}+b_{11}\sum X_8 X_{11}+b_{12}\sum X_8 X_{12}+b_{13}\sum X_8 X_{13}$$

$$(12)$$

10. $\sum X_9 Y=$
$$b_0\sum X_9+b_1\sum X_9 X_1+b_2\sum X_9 X_2+b_3\sum X_9 X_3+b_4\sum X_9 X_4+b_5\sum X_9 X_5+b_6\sum X_9 X_6+$$

$$+b_7\sum X_9 X_7+b_8\sum X_9 X_8+b_9\sum X_9^2+b_{10}\sum X_9 X_{10}+b_{11}\sum X_9 X_{11}+b_{12}\sum X_9 X_{12}+b_{13}\sum X_9 X_{13}$$

$$(13)$$

11. $\sum X_{10} Y=b_0\sum X_{10}+b_1\sum X_{10} X_1+b_2\sum X_{10} X_2+b_3\sum X_{10} X_3+b_4\sum X_{10} X_4+b_5\sum X_{10} X_5+b_6\sum X_{10} X_6+$

$$+b_7\sum X_{10} X_7+b_8\sum X_{10} X_8+b_9\sum X_{10} X_9+b_{10}\sum X_{10}^2+b_{11}\sum X_{10} X_{11}+b_{12}\sum X_{10} X_{12}+b_{13}\sum X_{10} X_{13}$$

$$(14)$$

12. $\sum X_{11} Y=b_0\sum X_{11}+b_1\sum X_{11} X_1+b_2\sum X_{11} X_2+b_3\sum X_{11} X_3+b_4\sum X_{11} X_4+b_5\sum X_{11} X_5+b_6\sum X_{11} X_6+$

$$+b_7\sum X_{11} X_7+b_8\sum X_{11} X_8+b_9\sum X_{11} X_9+b_{10}\sum X_{11} X_{10}+b_{11}\sum X_{11}^2+b_{12}\sum X_{11} X_{12}+b_{13}\sum X_{11} X_{13}$$

$$(15)$$

13. $\sum X_{12}Y=$

$b_0\sum X_{12}+b_1\sum X_{12}X_1+b_2\sum X_{12}X_2+b_3\sum X_{12}X_3+b_4\sum X_{12}X_4+b_5\sum X_{12}X_5+b_6\sum X_{12}X_6+$

$+b_7\sum X_{12}X_7+b_8\sum X_{12}X_8+b_9\sum X_{12}X_9+b_{10}\sum X_{12}X_{10}+b_{11}\sum X_{12}X_{11}+b_{12}\sum X_{12}^2+b_{13}\sum X_{12}X_{13}$

$$(16)$$

14. $\sum X_{13}Y=b_0\sum X_{13}+b_1\sum X_{13}X_1+b_2\sum X_{13}X_2+b_3\sum X_{13}X_3+b_4\sum X_{13}X_4+b_5\sum X_{13}X_5+b_6\sum X_{13}X_6+$

$+b_7\sum X_{13}X_7+b_8\sum X_{13}X_8+b_9\sum X_{13}X_9+b_{10}\sum X_{13}X_{10}+b_{11}\sum X_{13}X_{11}+b_{12}\sum X_{13}X_{12}+ b_{13}\sum X_{13}^2$

$$(17)$$

The computations of sums and means for the independent variables are given in Table 1.

And the computations of product sums for the independent and dependent variables are given in Table 2.

Table 1 The computation of sums and means for the independent variables

Independent variables	Total sum	Mean
$\sum X_1$	9678	
$\overline{X}_1$		52.885
$\sum X_2$	16836	
$\overline{X}_2$		92
$\sum X_3$	4649.50	
$\overline{X}_3$		25.407
$\sum X_4$	240	
$\overline{X}_4$		1.311
$\sum X_5$	14760	
$\overline{X}_5$		80.656
$\sum X_6$	270	
$\overline{X}_6$		1.475
$\sum X_7$	3960	
$\overline{X}_7$		21.639
$\sum X_8$	1080	
$\overline{X}_8$		5.902
$\sum X_9$	600	
$\overline{X}_9$		3.279
$\sum X_{10}$	1050	
$\overline{X}_{10}$		5.738
$\sum X_{11}$	540	
$\overline{X}_{11}$		2.951
$\sum X_{12}$	2430	
$\overline{X}_{12}$		13.279
$\sum X_{13}$	37947	
$\overline{X}_{13}$		207.361

Table 2 The computation of product sums for the independent variables

X_1	Value	X_2	Value	X_3	Value	X_4	Value
$\sum X_1 X_1$	1020036	$\sum X_2 X_2$	2059604	$\sum X_3 X_3$	403766.3	$\sum X_4 X_4$	936
$\sum X_1 X_2$	902327	$\sum X_2 X_1$	902327	$\sum X_3 X_1$	220561.5	$\sum X_4 X_1$	25545
$\sum X_1 X_3$	220561.5	$\sum X_2 X_3$	431498.5	$\sum X_3 X_2$	431498.5	$\sum X_4 X_2$	17591
$\sum X_1 X_4$	25545	$\sum X_2 X_4$	17591	$\sum X_3 X_4$	5249	$\sum X_4 X_3$	5249
$\sum X_1 X_5$	1539360	$\sum X_2 X_5$	1494360	$\sum X_3 X_5$	357480	$\sum X_4 X_5$	20520
$\sum X_1 X_6$	28830	$\sum X_2 X_6$	22650	$\sum X_3 X_6$	5460	$\sum X_4 X_6$	840
$\sum X_1 X_7$	420090	$\sum X_2 X_7$	362010	$\sum X_3 X_7$	90090	$\sum X_4 X_7$	14850
$\sum X_1 X_8$	116640	$\sum X_2 X_8$	66420	$\sum X_3 X_8$	25740	$\sum X_4 X_8$	2880
$\sum X_1 X_9$	60750	$\sum X_2 X_9$	50400	$\sum X_3 X_9$	14250	$\sum X_4 X_9$	2850
$\sum X_1 X_{10}$	112980	$\sum X_2 X_{10}$	56910	$\sum X_3 X_{10}$	23310	$\sum X_4 X_{10}$	3780
$\sum X_1 X_{11}$	57150	$\sum X_2 X_{11}$	60840	$\sum X_3 X_{11}$	11250	$\sum X_4 X_{11}$	1800
$\sum X_1 X_{12}$	254070	$\sum X_2 X_{12}$	293760	$\sum X_3 X_{12}$	49005	$\sum X_4 X_{12}$	9990
$\sum X_1 X_{13}$	4121696	$\sum X_2 X_{13}$	3663256	$\sum X_3 X_{13}$	866069	$\sum X_4 X_{13}$	98848

Table 2 (continue)

X_5	Value	X_6	Value	X_7	Value	X_8	Value
$\sum X_5 X_5$	5313600	$\sum X_6 X_6$	0	$\sum X_7 X_7$	1306800	$\sum X_8 X_8$	194400
$\sum X_5 X_1$	1539360	$\sum X_6 X_1$	28830	$\sum X_7 X_1$	420090	$\sum X_8 X_1$	116640
$\sum X_5 X_2$	1494360	$\sum X_6 X_2$	22650	$\sum X_7 X_2$	362010	$\sum X_8 X_2$	66420
$\sum X_5 X_3$	357480	$\sum X_6 X_3$	5460	$\sum X_7 X_3$	90090	$\sum X_8 X_3$	25740
$\sum X_5 X_4$	20520	$\sum X_6 X_4$	840	$\sum X_7 X_4$	14850	$\sum X_8 X_4$	2880
$\sum X_5 X_6$	0	$\sum X_6 X_5$	0	$\sum X_7 X_5$	0	$\sum X_8 X_5$	0
$\sum X_5 X_7$	0	$\sum X_6 X_7$	0	$\sum X_7 X_6$	0	$\sum X_8 X_6$	0
$\sum X_5 X_8$	0	$\sum X_6 X_8$	0	$\sum X_7 X_8$	0	$\sum X_8 X_7$	0
$\sum X_5 X_9$	0	$\sum X_6 X_9$	0	$\sum X_7 X_9$	0	$\sum X_8 X_9$	0
$\sum X_5 X_{10}$	0	$\sum X_6 X_{10}$	0	$\sum X_7 X_{10}$	0	$\sum X_8 X_{10}$	0
$\sum X_5 X_{11}$	0	$\sum X_6 X_{11}$	0	$\sum X_7 X_{11}$	0	$\sum X_8 X_{11}$	0
$\sum X_5 X_{12}$	0	$\sum X_6 X_{12}$	0	$\sum X_7 X_{12}$	0	$\sum X_8 X_{12}$	0
$\sum X_5 X_{13}$	6042600	$\sum X_6 X_{13}$	112050	$\sum X_7 X_{13}$	1650330	$\sum X_8 X_{13}$	451800

Table 2 (continue)

X_9	Value	X_{10}	Value	X_{11}	Value	X_12	Value	X_{13}	Value
$\sum X_9 X_9$	90000	$\sum X_{10} X_{10}$	220500	$\sum X_{11} X_{11}$	48600	$\sum X_{12} X_{12}$	656100	$\sum X_{13} X_{13}$	15699219
$\sum X_9 X_1$	60750	$\sum X_{10} X_1$	112980	$\sum X_{11} X_1$	57150	$\sum X_{12} X_1$	254070	$\sum X_{13} X_1$	4121696
$\sum X_9 X_2$	50400	$\sum X_{10} X_2$	56910	$\sum X_{11} X_2$	60840	$\sum X_{12} X_2$	293760	$\sum X_{13} X_2$	3663256
$\sum X_9 X_3$	14250	$\sum X_{10} X_3$	23310	$\sum X_{11} X_3$	11250	$\sum X_{12} X_3$	49005	$\sum X_{13} X_3$	866069
$\sum X_9 X_4$	2850	$\sum X_{10} X_4$	3780	$\sum X_{11} X_4$	1800	$\sum X_{12} X_4$	9990	$\sum X_{13} X_4$	98848
$\sum X_9 X_5$	0	$\sum X_{10} X_5$	0	$\sum X_{11} X_5$	0	$\sum X_{12} X_5$	0	$\sum X_{13} X_5$	6042600
$\sum X_9 X_6$	0	$\sum X_{10} X_6$	0	$\sum X_{11} X_6$	0	$\sum X_{12} X_6$	0	$\sum X_{13} X_6$	112050
$\sum X_9 X_7$	0	$\sum X_{10} X_7$	0	$\sum X_{11} X_7$	0	$\sum X_{12} X_7$	0	$\sum X_{13} X_7$	1650330
$\sum X_9 X_8$	0	$\sum X_{10} X_8$	0	$\sum X_{11} X8$	0	$\sum X_{12} X_8$	0	$\sum X_{13} X_8$	451800
$\sum X_9 X_{10}$	0	$\sum X_{10} X_9$	0	$\sum X_{11} X_9$	0	$\sum X^{12} X_9$	0	$\sum X_{13} X_9$	232650
$\sum X_9 X_{11}$	0	$\sum X_{10} X_{11}$	0	$\sum X_{11} X_{10}$	0	$\sum X_{12} X_{10}$	0	$\sum X_{13} X_{10}$	433650
$\sum X_9 X_{12}$	0	$\sum X_{10} X_{12}$	0	$\sum X_{11} X_{12}$	0	$\sum X_{12} X_{11}$	0	$\sum X_{13} X_{11}$	227250
$\sum X_9 X_{13}$	232650	$\sum X_{10} X_{13}$	433650	$\sum X_{11} X_{13}$	227250	$\sum X_{12} X_{13}$	1019250	$\sum X_{13} X_{12}$	1019250

In Table 3 shown the computation of product sums for the dependent variables .

Table 3 The computation of product sums for the dependent variables

Dependent variables	Values
$\sum Y$	1272.307
$\bar{Y}$	6.952
$\sum X_1 Y$	134085.06
$\sum X_2 Y$	114985.688
$\sum X_3 Y$	29012.105
$\sum X_4 Y$	3374.32
$\sum X_5 Y$	202374.72
$\sum X_6 Y$	3707.25
$\sum X_7 Y$	53641.50
$\sum X_8 Y$	16250.40
$\sum X_9 Y$	8670
$\sum X_{10} Y$	15608.25
$\sum X_{11} Y$	7258.50
$\sum X_{12} Y$	32663.25
$\sum X_{13} Y$	524157.73

With account of calculated values for the independent and dependent variables which are shown in the Tables 1,2 and 3 ,we have the following system which consists from the fourteen normal equations for solving and calculation of their coefficients:

$1272.307 = 183b_0 + 9678b_1 + 16836b_2 + 4649.5b_3 + 240b_4 + 14760b_5 + 270b_6 + 3960b_7 + 1080b_8 + 600b_9 +$
$+1050b_{10} + 540b_{11} + 2430b_{12} + 37947b_{13};$

$134085.06 = 9678b_0 + 1020036b_1 + 902327b_2 + 220561.50b_3 + 25545b_4 + 1539360b_5 28830b_6$
$+420090b_7 + 116640b_8 + 60750b_9 + 112980b_{10} + 57150b_{11} + 254070b_{12} + 4121696b_{13};$

$114985.688 = 16836b_0 + 902327b_1 + 2059604b_2 + 431498.5b_3 + 17591b_4 + 1494360b_5 + 22650b_6 +$
$+362010b_7 + 66420b_8 + 50400b_9 + 56910b_{10} + 60840b_{11} + 293760b_{12} + 3663256b_{13};$

$29012.105 = 4649.5b_0 + 220561.5b_1 + 431498.5b_2 + 403766.25b_3 + 5249b_4 + 357480b_5 + 5460b_6 +$
$+90090b_7 + 25740b_8 + 14250b_9 + 23310b_{10} + 11250b_{11} + 49005b_{12} + 866069b_{13};$

$3374.32 = 240b_0 + 25545b_1 + 17591b_2 + 5249b_3 + 936b_4 + 20520b_5 + 840b_6 + 14850b_7 + 2880b_8 +$
$+2850b_9 + 3780b_{10} + 1800b_{11} + 9990b_{12} + 98848b_{13};$

$202374.72 = 14760b_0 + 1539360b_1 + 1494360b_2 + 357480b_3 + 20520b_4 + 5313600b_5 + 6042600b_{13};$

$3707.25 = 270b_0 + 28830b_1 + 22650b_2 + 5460b_3 + 840b_4 + 112050b_{13};$

$53641.50 = 3960b_0 + 420090b_1 + 362010b_2 + 90090b_3 + 14850b_4 + 1306800b_7 + 1650330b_{13};$

$16250.40 = 1080b_0 + 116640b_1 + 66420b_2 + 25740b_3 + 2880b_4 + 194400b_8 + 451800b_{13};$

$8670 = 600b_0 + 60750b_1 + 50400b_2 + 14250b_3 + 2850b_4 + 90000b_9 + 232650b_{13};$

$15608.25 = 1050b_0 + 112980b_1 + 56910b_2 + 23310b_3 + 3780b_4 + 220500b_{10} + 433650b_{13};$

$7258.50 = 540b_0 + 57150b_1 + 60480b_2 + 11250b_3 + 1800b_4 + 48600b_{11} + 227250b_{13};$

$32663.25 = 2430b_0 + 254070b_1 + 293760b_2 + 49005b_3 + 9990b_4 + 656100b_{12} + 1019250b_{13};$

$524157.73 = 37947b_0 + 4121696b_1 + 3663256b_2 + 866069b_3 + 98848b_4 + 6042600b_5 + 112050b_6 +$
$+1650330b_7 + 451800b_8 + 232650b_9 + 433650b_{10} + 227250b_{11} + 1019250b_{12} + 15699219b_{13}$

Computations of these fourteen normal equations gave to us the possibility to find the unknown coefficient of correlation which are equal:

$b_0=1.357$; $b_1=0.1615$; $b_2=-0.0101$; $b_3=-0.0002$; $b_4=0.5686$; $b_5=0.0016$; $b_6=0.0106$;

$b_7=0.0045$; $b_8=0.0018$; $b_9=-0.0041$; $b_{10}=0.0006$; $b_{11}=-0.1241$; $b_{12}=-0.0157$; $b_{13}=-0.0117$

So, the multiple regression model for the ship speed in the tropics and wave sea is :

$$\hat{Y} = 1.357 + 0.1615X_1 - 0.0101X_2 - 0.0002X_3 + 0.5686X_4 + 0.0016X_5 + 0.0106X_6 +$$
$$+ 0.0045X_7 + 0.0018X_8 - 0.0041X_9 + 0.0006X_{10} - 0.1241 X_{11} - 0.0157X_{12} - 0.0117X_{13}$$

$$(18)$$

5.2 Analysis of empirical formula for the evaluation of ship speed in the tropics and wave sea

The data of analysis of empirical formula (18) for the evaluation of ship speed in the tropics and
*wave sea is given in **Appendix 2**.*
In base of evaluation of empirical formula are put such parameters as :

a) Multiple correlation coefficient which is equal:

$$R^2 = (\sum \hat{Y}Y) / [(\sum \hat{Y}^2) (Y^2)] = \sum \hat{Y}^2 / \sum Y^2 \qquad (19)$$

At $\sum \hat{Y}^2 = 13165.0015$ and $\sum Y^2 = 13413.8171$ we have $R^2 = 0.9815$; $R = 0.991$

So, the calculation of multiple correlation coefficient $R = 0.991 < 1.0$ shows that 99.10 percent of the variability in Y. This is a good relation ship between Y_i and X_i.

b) Variance which is equal

$$S_{y/x}^2 = \sum (Y_i - \hat{Y}_i)^2 / n\text{-}k\text{-}1 \qquad (20)$$

At n=183 (observation data) then k+1=14 and $S_{y/x}^2 = 828.752/169 = 4.904$; $S_{y/x} = 2.214$

Conclusions

1. In this paper ,the multiple regression analysis is presented for evaluation of ship speed in the tropics and wave sea;

2. This empirical formula will give the possibility in the future for the operational people indirectly and approximately to predict the optimal ship speed at the different conditions of sailing ship in the tropics and wave sea ,particularly for the merchant vessels and oil tankers.

CHAPTER 6

Prediction of ship speed and some parameters in the tropics

6.1 Non-linear correlation between the ship of speed and duration in-service

The dependence of ship speed from its duration in-service was described in the above-named works. But now, in Table 1 is presented the new observation data which will give the possibility to define in detail the forecasting process .

The main goal of these data to show the changing of monthly ship speed by methods of mathematical statistics in dependence from the running period of this ship in the tropics and besides to predict this ship speed on the next navigation period.

It s known that forecasting method is one of the most important function which is performed by many researchers today. In our conditions is very useful tool that to know the value of ship speed on the future period of its running that successfully to plan the operational processes in question of transportation by sea the different cargoes to the countries having the tropical climate and also to maintenance in good conditions the main marine diesel engine and ship in whole .

Table 1 Observation monthly data of ship speed in the tropics

Month	Duration in-service N ,days	Average ship speed Vk, mi/h
August	30	14.916
September	60	14.814
October	90	14.12
November	120	12.775
December	150	13.073
January	180	12.974

In Figure 1.1 is given a scatter diagram showing the relationship between the duration in-service and ship speed.

As we see from Figure 1.1 that the relationship between duration in-service and ship speed departs from linearity, i.e this function has the nonlinear correlation. So, with increasing of duration in-service, the ship speed decreases considerably.

As was shown in the previous works we see that the main reason of such dependence there is a marine growth (fouling).

However, there are many research works which are made by the other authors [1] ,[2] and [3] in question of prevention the body of ship from fouling. Unfortunately, all these methods and arrangements do not guarantee the full success because they only decreases the value of fouling the ship's hull and author of this work admits that practically to avoid the fouling impossible and particularly for the ships running in the tropics.

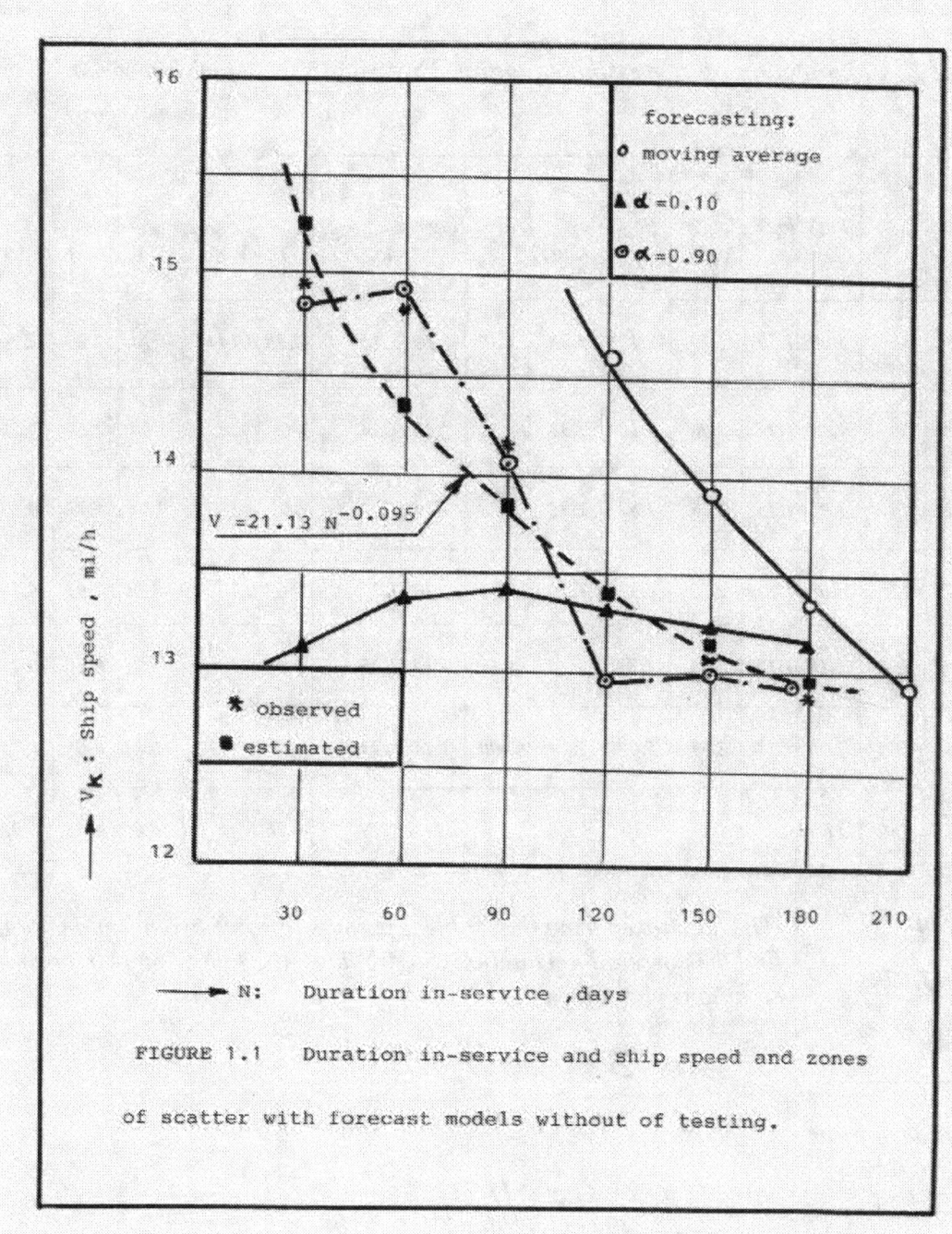

FIGURE 1.1   Duration in-service and ship speed and zones

of scatter with forecast models without of testing.

And his conclusions and arguments the author shows on the examples of merchant ship which had the six-month navigation in the tropical seawaters.

Anatoly Rozenblat

In Table 2 is given the computation of values, on the basis of data shown in Table 1, which are used in determining the logarithmic measures of relationship between duration in-service and ship speed.

Table 2 Computation of regression equation $V_k = 21.13N^{-0.095}$ for ship speed

Y	X	logX	logY	(logX)x(logY)	$(logX)^2$	$(logY)^2$	$\hat{Y}c$	(logYc)	$(logYc)^2$	$Y-\hat{Y}c$	$(Y-\hat{Y}c)^2$	$Y-\bar{Y}$	$(Y-\bar{Y})^2$	$(Yc-\bar{Y})^2$
14.92	30	1.48	1.17	1.73	2.18	1.38	15.30	1.19	1.40	-0.38	0.144	1.14	1.29	2.298
14.81	60	1.78	1.17	2.08	3.16	1.37	14.32	1.16	1.34	0.49	0.243	1.04	1.07	0.294
14.12	90	1.95	1.15	2.25	3.82	1.32	13.78	1.14	1.29	0.34	0.12	0.34	0.12	0
12.78	120	2.08	1.11	2.29	4.32	1.22	13.41	1.13	1.27	-0.63	0.40	-1.00	1.01	0.14
13.07	150	2.18	1.12	2.43	4.73	1.25	13.13	1.12	1.25	-0.05	0	-0.71	0.49	0.43
12.97	180	2.26	1.11	2.51	5.09	1.24	12.90	1.11	1.23	0.07	0	-0.81	0.65	0.77
Total 82.67	630	11.72	6.83	13.29	23.30	7.78		6.84	7.79	-0.16	0.91	-0.06	4.63	3.92

Mean:
$\bar{Y}=13.779$;
$\bar{X}=105$

Standard error
$\sigma=0.045$

Coefficient of determination
$R^2=0.80$

The data taken from this Table 2 permits to solve two normal equation find the correlation equation .So ,we have the following two normal equations view of:

1. $\sum logY = n\,loga + b\sum logX$

2. $\sum(logX \cdot logY) = loga \sum logX + b\sum(logX)^2$

   $6.827 = 6\,loga + 11.719$
   $13.2297 = 11.719\,loga + 23.301b$

The solving these two normal equations give to us the following regres equation: $logYc = 1.325 - 0.095\,logX$ or $log\,V_k = 1.325 - 0.095\,logN$

So, the ship speed is equal $V_k = 21.13N^{-0.095}$ (1) and is the function of

Correlation coefficient *r=0.894*	*duration in-service (N).*

Analyzing the Figure 1.1 and Table 2 we can now to determine the statistical parameters

of this nonlinear regression equation view of $V_k=21.13N^{-0.095}$:

1. *Standard error estimates by formula* $\sigma^2 = [\ \Sigma(\log Y)^2 - \Sigma(\log \hat{Y}_c)^2\ ]\ /\ n = \Sigma(\log \hat{Y}_c)^2\ /n$ *where n= number of observations (n=6) and* $\sigma=0.045$;

2. *Coefficient of determination is equal* $R = [\ \Sigma(Y-\bar{Y})^{-2} - \Sigma(Y-\hat{Y}_c)^2\ /\ [\ \Sigma(Y-\bar{Y})^{-2}$, $R=0.80$

Thus, the regression equation $Y_c=21.13X^{-0.095}$ *accounts for 80 percent of the variability in* Y_i, *indicating that a very strong nonlinear relationship has been identified in the above*
Indicated equation for the ship speed (V_k) in dependence of in-service (N) .

3. *Correlation coefficient (r) is equal* $r = (R^2)^{0.5}$, $r=0.894$

6.2 Forecasting models and their compare in evaluation of ship speed in the tropics

A. Moving average forecast

Primary we will consider "the moving average forecasts" and later the " exponential smoothing time series " models which rather well are used for the forecasting of ship speed.
  In Table 3 is shown three-month moving average ship speed forecasting and these data are indicated in Figure 1.1.

Table 3 Three-moving ship speed in the tropics

Month	Duration in-service N,days	Most recent average three-month speed , ΣV_k	Average forecasting ship speed, V_k^F
August	30		-
Septembe	60		-
October	90		-

November	120	14.916+14.814+14.12	14.617
December	150	14.814+14.12+12.775	13.903
January	180	14.12+12.775+13.073	13.323
February	210	12.775+13.073+12.974	12.941

From Table 3 we see that the forecasting ship speed decreases with duration in-service of this ship in the tropics. Analyzing of data shown in Table 3 we see that the forecasting ship speed decreases, but which view of regression model has this dependence we do not know.

With objective of determining this regression equation we must make some important steps such as:

1. *To evaluate the data of Table 3,primary accepting into attention that these data*

has the linear regression model view of Y=b_0+b_1X. On the basis of these conclusions are made the calculations and data of forecasting ship speed which are shown in Table 4.

F

Table 4 Evaluation of forecasting ship speed V_k for the linear regression model

Y	X	X^2	Y^2	XY	$\bar{Y_c}$	$Y-\bar{Y}$	$(Y-\bar{Y})^2$	$(\hat{Y_c}-\bar{Y})$	$(Y-\hat{Y_c})$	$(\hat{Y_c}-\bar{Y})^2$	$(Y-\hat{Y_c})^2$	$X-\bar{X}$	$(X-\bar{X})^2$	$(X-\bar{X})(Y-\bar{Y})$
14.62	120	14400	213.7	1754.0	14.5	0.92	0.85	0.85	0.07	0.731	0.0004	-45	2025	41.445
13.90	150	22500	193.3	2085.5	13.9	0.21	0.04	0.29	-0.08	0.081	0.0006	-15	225	-3.105
13.32	180	32400	177.5	2398.1	13.4	-0.4	0.14	-0.29	-0.09	0.081	0.0007	15	225	-5.595
12.94	210	44100	167.5	2717.6	12.8	-0.8	0.57	-0.86	0.10	0.731	0.010	45	2025	33.975
Total 54.78	660	113400	751.9	8955.2		0	1.60	0	0	1.624	0.0117	0	4500	66.72

Mean: $\bar{Y}$=13.696 ; $\bar{X}$=165 Variance $S_{y/x}$=0.73	So, from the Table 4 we can calculate the coefficients b_0 and b_1 of linear regression model $Y=b_0+b_1X$ where $b_1=[(\sum X_i Y_i) - (\sum X_i)(\sum Y_i)/n]/[(X_i)^2 - (\sum X_i)^2/n]$ and $b_1=$ -

Coefficient of determination $R^2 = 0.993$	**0.019** $b_0 = \bar{Y} - b_1\bar{X}$ and $b_0 = 16.831$ *And the linear regression model has view of :*
Coefficient of correlation $r = 0.786$	$\hat{Y} = 16.831 - 0.019X$ **(2)** *So, the forecasting ship speed is equal* $V_k^F = 16.831 - 0.019N$

Statistical parameters of this linear regression model view of $\hat{Y} = 16.831 - 0.019X$ *can be evaluated in the following steps:*

1.Pearson's product-moment correlation coefficient is equal

$$r = [\ \Sigma(X_i - \bar{X})(Y_i - \bar{Y})\ ] \ / \{\ [(\Sigma(X_i - \bar{X})^{-2}]^{0.50} \cdot [\ \Sigma(Y_i - \bar{Y})^{-2}]^{0.50}\ \} \ \text{ where } \ r = 0.786$$

2.Coefficient of determination is equal

$$R^2 = (\ SST - SSE)/SST \ \text{ where } \ SST = \Sigma(Y - \bar{Y})^2 = 1.60; \ SSE = \Sigma(Y - Y_c)^{\wedge 2} = 0.0117 \ and$$
$$R = 0.993$$

3.Sum of squared deviations is equal $S_{y/x} = [\ \Sigma(Y_i - \bar{Y})^2\ ]/n - 1$ *; where* $S_{y/x} = 0.73.$

So ,we determined the statistical parameters for the forecasting ship speed linear model of view $\hat{Y} = 16.831 - 0.019X$ *or* $V_k^F = 16.831 - 0.019N$ *which is shown in Figure 1.2.*

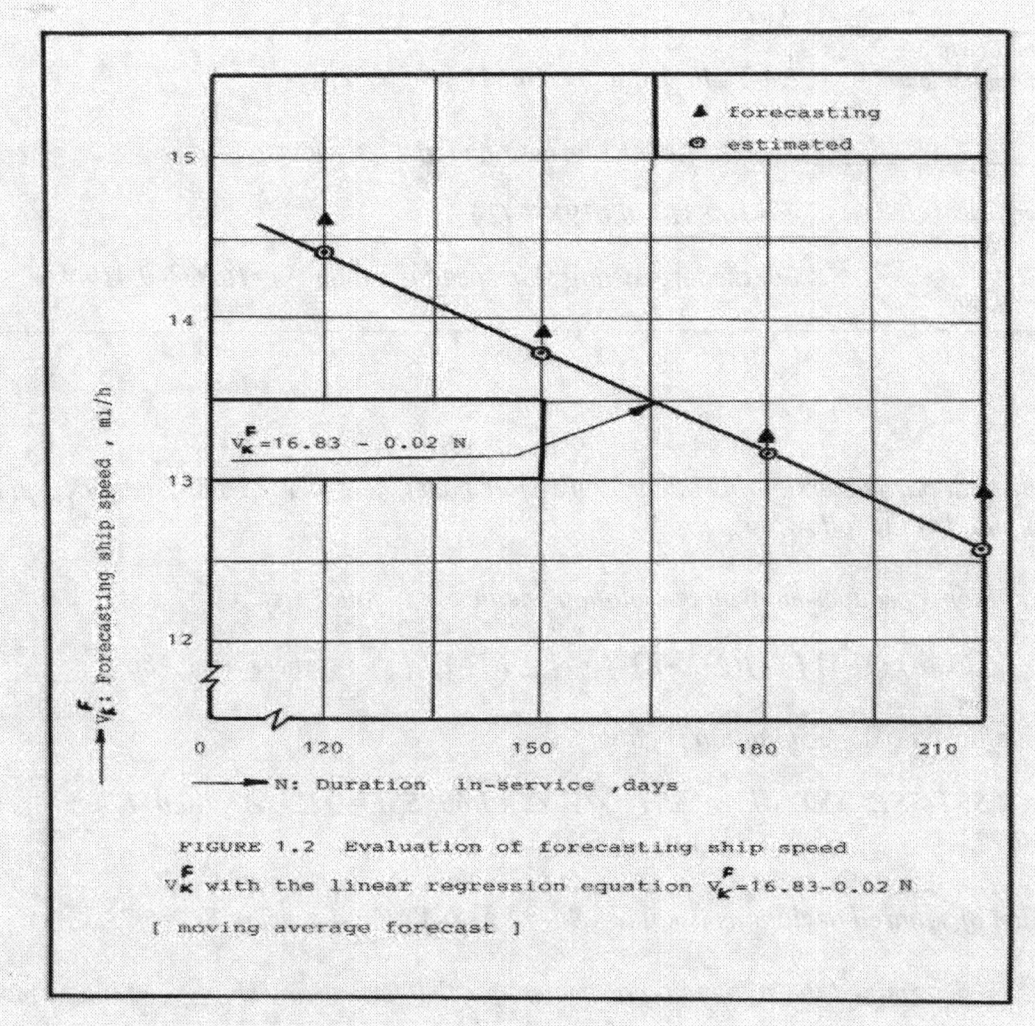

FIGURE 1.2 Evaluation of forecasting ship speed V_k^f with the linear regression equation $V_k^f = 16.83 - 0.02 \, N$

[moving average forecast]

And now that we could compare the forecasting models in evaluation of ship speed in the tropics, necessary to calculate the other forecasting model and determine the view of its regression equation and character of this functional dependency.

In Table 5 is shown the data for evaluation of forecasting (moving average model) ship (V_k^F) for the nonlinear regression equation view of $V_k^F = 44.57 N^{-0.235}$.

Table 5 Evaluation of forecasting ship speed V_k^F for the nonlinear regression equation

Y	X	logX	logY	logXlogY	$(logX)^2$	$(logY)^2$	$\hat{Y}_c$	logYc	$(logYc)^2$	$Y-\bar{Y}$	$Y-\hat{Y}_c$	$(Y-Y)^{-2}$	$(Y-Yc)^{\wedge 2}$

14.62	120	2.079	1.165	2.422	4.322	1.357	14.5	1.16	1.35	0.92	0.13	0.85	0.02
13.90	150	2.176	1.117	2.431	4.735	1.248	13.7	1.14	1.29	0.21	0.18	0.04	0.03
13.32	180	2.255	1.125	2.537	5.085	1.266	13.1	1.12	1.25	-0.37	0.18	0.14	0.03
12.94	210	2.322	1.112	2.582	5.392	1.237	12.7	1.10	1.22	-0.76	0.24	0.57	0.06
Total 54.78	660	8.832	4.519	9.972	19.534	5.108		4.52	5.113	0	0.72	1.60	0.14

Mean: $\overline{Y}=13.696; \overline{X}=165$	The data is taken from Table 5 permits to solve two normal equations and find the regression equation. So ,we have the following two equations view of: 1. $\sum logY=nloga+b\sum logX$ 2. $\sum(logXlogY)=loga\sum log+b\sum(logX)^2$ $4.519=4loga+b(8.832)$ $9.972=loga(8.832)+b(19.534)$
Standard error $\sigma=1.131$	The solving these two normal equations we have $loga=1.649; b=-0.235$ and regression equation has view $logY_c=1.649-0.235logX.$
Coefficient of determination $R^2=0.915$	So, the forecasting ship speed (V_k) is equal $V_k=44.57N^{-0.235}$ (3) and we see that this ship speed is also the function of duration in-service.
Correlation Coefficient $r=0.957$	

Analyzing the data from Table 5 and regression equation view of $V_k=44.57N^{-0.235}$ we can determine the statistical parameters of this nonlinear regression model:

1.Standard error is calculated by formula $\sigma=\sum(logY_c)^2/n =[\sum(logY)^2 -\sum(logY_c)^2] /n$;

at $n=4$ $\sum(logY_c)^2 =5.113$ we have $\sigma=1.131$

2.Coefficient of determination is equal $R^2 = [\sum(Y-\overline{Y})^2 - \sum(Y-Y_c)^2] /[\sum(Y-\overline{Y})^2] ; R^2=0.915$

Thus ,the regression equation $\hat{Y}_c=44.57N^{-0.235}$ accounts for 91.50 percent of the variability in Y_i, indicating that a very strong nonlinear relationship has been identified in the above-named equation for the forecasting ship speed in dependence of in-service (N) .This functional model is shown in Figure 1.3.

3. Correlation coefficient is equal $r = (R^2)^{0.50}$, $r = 0.957$

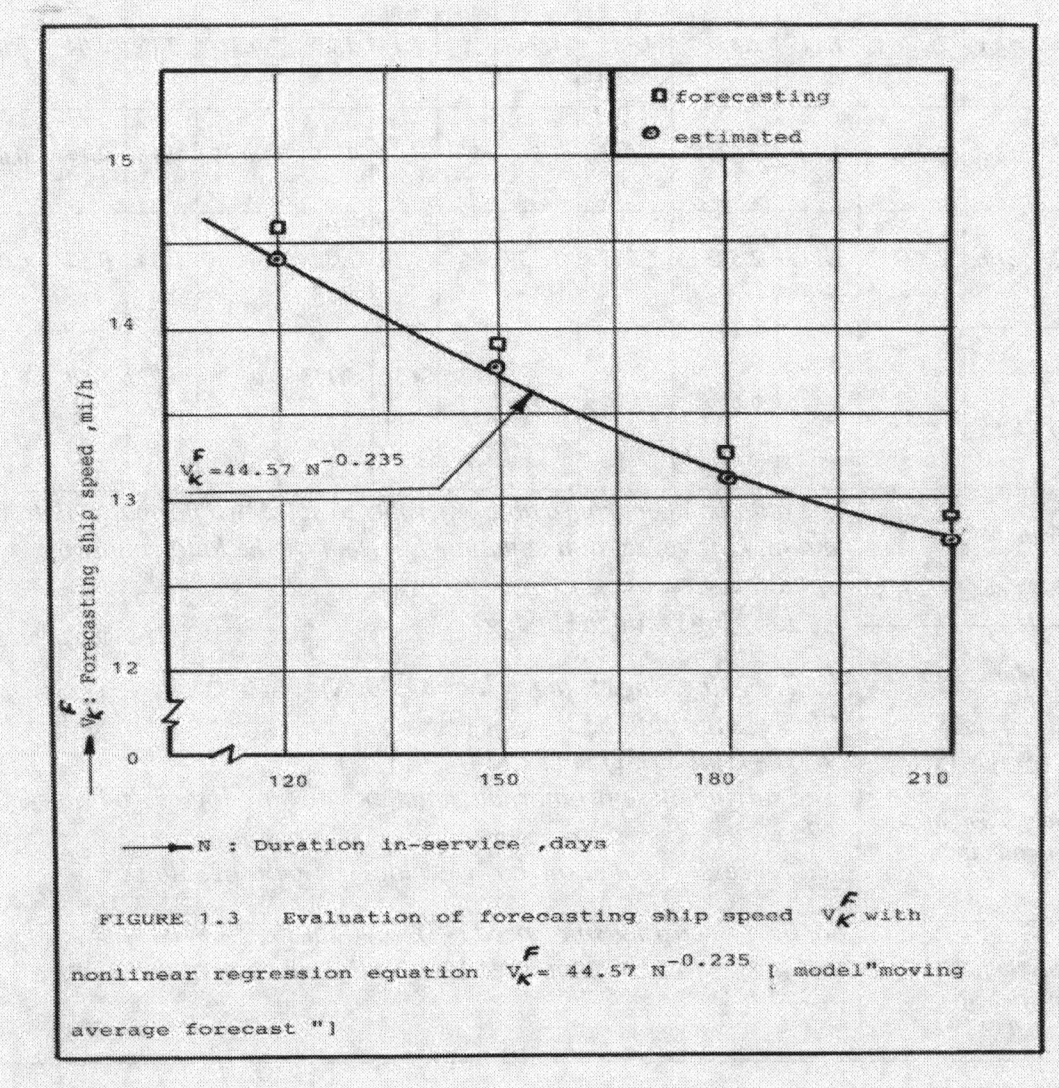

FIGURE 1.3   Evaluation of forecasting ship speed V_k^F with nonlinear regression equation $V_k^F = 44.57 \ N^{-0.235}$ [model"moving average forecast "]

So ,for comparison of the best forecasting regression model in question of determining the correct type of equation , use the well-known equations which must show the minimum value in our case ,i.e we have the following equations view of :

$$\text{min } MAD = \Sigma |Y - \hat{Y_c}| / n \text{ and } \text{min } MSE = \Sigma (Y - \hat{Y_c})^2 / n$$

In Table 6 is shown the general parameters of evaluation the forecasting ship speed (V_k^F) with **using method " Moving average forecasts".**

Table 6 Evaluation and comparison of method "Moving average forecast" for the forecasting ship speed.

Linear model	Nonlinear model	Results and conclusions
Regression equation $\hat{Y} = 16.831 - 0.019X$	**Regression equation** $\hat{Y} = 44.57X^{-0.235}$	*Comparing and analyzing two regression models, we accept that forecasting ship speed submits the better to the linear regression model which is shown in Figure 1.1 and has equation view* *of* $\hat{Y_c} = 16.831 - 0.019X$ *or* $V_k^F = 16.831 - 0.019N$
1.Mean : $\bar{Y} = 13.696$; $\bar{X} = 165$	*1.Mean :* $\bar{Y} = 13.696$; $\bar{X} = 165$	
2.Coefficient of correlation $r = 0.786$	*2.Coefficient of correlation* $r = 0.957$	
3.Coefficient of determination $R^2 = 0.993$	*3.Coefficient of determination* $R^2 = 0.915$	
4.Variance $S_{y/x} = 0.73$	*4.Standard error* $\sigma = 1.131$	
5. min MAD $= \sum \lvert Y - \hat{Y_c} \rvert / n$ *min MAD* $= 0/4 = 0$	*5. min MAD* $= \sum \lvert Y - \hat{Y_c} \rvert / n$ *min MAD* $= 0.721/4 = 0.180$	
6. min MSE $= \sum (Y - \hat{Y_c})^2 / n$ *min MSE* $= 0.0117/4 = 0.003$	*6.min MSE* $= \sum (Y - \hat{Y_c})^2 / n$ *min MSE* $= 0.136/4 = 0.034$	

B. The exponential smoothing time series model for forecasting of ship speed in the tropics

In accordance with this model, we can make choose "the seed" forecast which can be based primary on a "best guess". In our case, we determine this value from the data of Table 3, where the forecasting ship speed of this "seed" is equal $V_k = 12.941 \sim 13.00$ *mi/h* .

Using the data which is given in Table 1 , we can calculate the single exponential forecast for the ship speed at the different smoothing constant α which is determined by two values equal $\alpha = 0.10$ and $\alpha = 0.90$.

In Table 7 is given the single exponential forecast method and calculations of the ship speed with value of $\alpha=0.10$ and in Table 8 are made the calculations for the forecasting of the ship speed with value of $\alpha=0.90$.

Table 7 Single exponential forecast for the ship speed at smoothing constant $\alpha=0.10$

Month	Duration in-service, days	Average ship speed, mi/h	Smoothing constant $\alpha=0.10$	
			$\alpha(Y_t - \hat{Y}_t)$	Forecasting ship speed $\hat{Y}_{t+1}$
August	30	14.916	0.192	13.192
September	60	14.814	0.162	13.354
October	90	14.12	0.077	13.431
November	120	12.775	-0.066	13.365
December	150	13.073	- 0.029	13.336
January	180	12.974	-0.036	13.30

Analyzing the data from Table 7 ,we see that the forecasting ship speed has the variable character and its value in January can have equal as $V_k=13.30$ *mi/h* ,i.e this value is more than actual statistical value at the same month as $V_k=12.974$ *mi/h*.

Table 8 Single exponential forecast for the ship at smoothing constant $\alpha=0.90$

Month	Duration in-service, days	Average ship speed, mi/h	Smoothing constant $\alpha=0.90$	
			$\alpha(Y_t - \hat{Y}_t)$	Forecasting ship speed $\hat{Y}_{t+1}$

August	*30*	*14.916*	*1.724*	*14.724*
September	*60*	*14.814*	*0.081*	*14.805*
October	*90*	*14.12*	*- 0.617*	*14.188*
November	*120*	*12.775*	*-1.272*	*12.916*
December	*150*	*13.073*	*0.141*	*13.057*
January	*180*	*12.974*	*- 0.075*	*12.982*

Analysis of Table 8 shows that the forecasting ship speed in January month with smoothing constant $\alpha=0.90$ is less than in the same month with the smoothing constant $\alpha=0.10$.

In Figure 1.4 is shown the evaluation of forecasting ship speed with the nonlinear regression equation $V_k=13.40N^{-0.0012}$ **[model" single exponential forecast" with smoothing constant $\alpha=0.10$].**

Analysis of Figure 1.4 shows that with increasing of duration in-service ,the forecasting ship speed for the above-named model considerably decreases.

In Table 9 is given the computation of regression equation $V_k= 13.40N^{-0.0012}$ *and also the statistical parameters of this regression model.*

With objective of determination the best view of regression equation and statistical parameters of the forecasting ship speed, for the model" single exponential forecast" ,we must test the other regression models of this functional dependence.

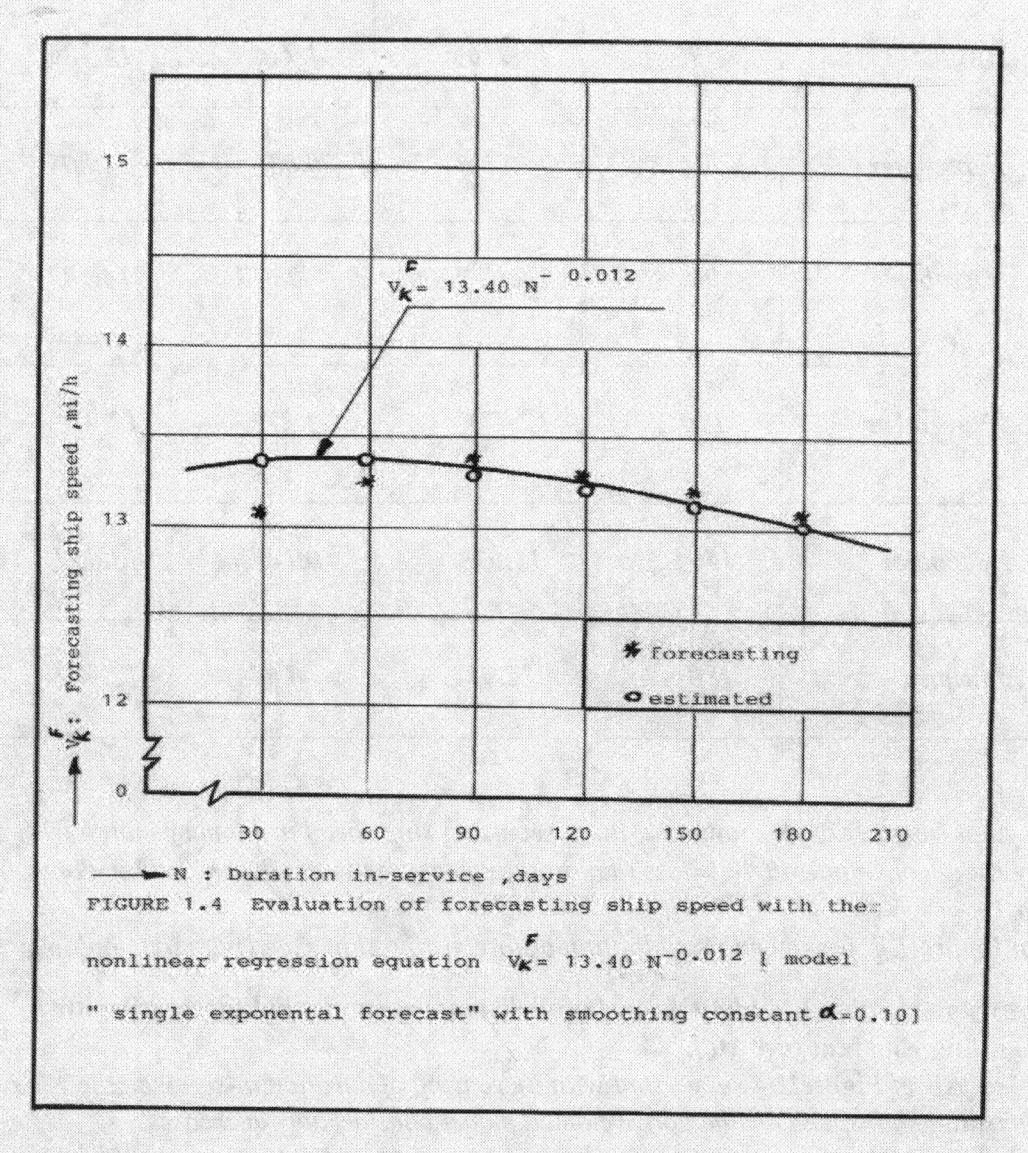

FIGURE 1.4 Evaluation of forecasting ship speed with the nonlinear regression equation $V_K^F = 13.40\ N^{-0.012}$ [model " single exponental forecast" with smoothing constant $\alpha = 0.10$]

Statistical parameters of the forecasting nonlinear regression equation $\hat{Y_c} = 13.40 X^{-0.0012}$:

1. Standard error of estimate $\qquad \sigma^2 = \sum (\log \hat{Y_c})^2 / n$, where $\sigma = 1.125$;

2. Coefficient of determination $\ R^2 = [\ \sum (\log \hat{Y_c})^2\ ] / [\ \sum (\log Y)^2\ ]$, where $R^2 = 0.999 \sim 1.0$;

3. Coefficient of correlation $\ r = (R^2)^{0.50}$,where $r = 0.999 \sim 1.0.$

Table 9 Evaluation of forecasting ship speed V_k^F for the nonlinear regression equation $\hat{Y}_c = 13.40\,X^{-0.0012}$ with single exponential smoothing constant $\alpha = 0.10$

Y	X	logX	logY	(logY)(logX)	$(logX)^2$	$(logY)^2$	$\hat{Y}_c$	$\widehat{(Y-Y_c)}$	$\widehat{(Y-Y_c)^2}$	$\overline{(Y-\bar{Y})^2}$	$\widehat{(logY_c)^2}$
13.192	30	1.477	1.120	1.654	2.182	1.255	13.3	-0.15	0.023	0.019	1.266
13.354	60	1.778	1.126	2.002	3.161	1.267	13.3	0.20	0.000	0.001	1.266
13.431	90	1.954	1.128	2.204	3.818	1.273	13.3	0.10	0.011	0.010	1.265
13.365	120	2.079	1.126	2.341	4.322	1.268	13.3	0.04	0.002	0.001	1.265
13.336	150	2.176	1.125	2.448	4.735	1.266	13.3	0.02	0.000	0.000	1.264
13.30	180	2.255	1.124	2.535	5.086	1.263	13.3	-0.02	0.000	0.001	1.264
Total: 79.97	630	11.719	6.749	13.184	23.303	7.592		0.01	0.037	0.032	7.590

Mean: $\bar{Y}=13.329$; $\bar{X}=105$	*The two normal equations for calculation of coefficients **b** and **loga**:* $$\sum LogY = nloga + blogX$$ $$\sum (logXlogY) = loga\sum logX + b\sum(logX)^2$$
Standard error $\sigma = 1.125$	
Coefficient of determination $R^2 = 1.00$	$6.749 = 6loga + 11.719b$ $13.184 = loga(11.719) + b(23.303)$

Correlation coefficient r=1.00	Solving these two normal equations we have **b=-0.0012; loga=1.127** and then the regression equation has view of $\log \hat{Y_c} = \log a + b \log X = 1.127 - 0.0012 \log X$; $\hat{Y_c} = 13.40 X^{-0.0012}$

In Table 10 is given the computation data for the single exponential forecasting ship speed (V_k) for the linear regression model with smoothing constant α=0.10.

Table 10 Evaluation of single exponential forecasting ship speed V_{kF} for the linear regression equation Y_c=13.287-0.0004X with the smoothing constant α=0.10

Y	X	X^2	Y^2	XY	$\hat{Y_c}$	$Y-\bar{Y}$	$(Y-\bar{Y})^2$	$(Y-\hat{Y_c})^2$	$(X-\bar{X})$	$(X-\bar{X})^2$	$(X-\bar{X})(Y-\bar{Y})$
13.19	30	900	174.03	395.8	13.3	-0.14	0.012	0.007	-75	5625	10.275
13.35	60	3600	178.33	801.2	13.3	0.03	0.001	0.008	-45	2025	-1.125
13.43	90	8100	180.39	1208.8	13.3	0.10	0.010	0.032	-15	225	-1.530
13.37	120	14400	178.62	1603.8	13.2	0.04	0.001	0.016	15	225	0.540
13.34	150	22500	177.85	2000.4	13.2	0.01	0.000	0.012	45	2025	0.315
13.30	150	32400	176.89	2394	13.2	-0.03	0.001	0.007	75	5625	-2.175
Total: 79.98	630	81900	1066.11	8404		0.004	0.032	0.082	0	15750	6.300

Mean: $\bar{Y}$ =13.329 ; $\bar{X}$ =105	Formula for determination of linear regression model has view $Y_c = b_0 + b_1 X$ and coefficients of b_0 and b_1 can be calculated by formula :
Correlation coefficient r= 0.28	$b_1 = \{ \sum X_i Y_i - [(\sum X_i)(\sum Y_i)/n] \}/\{[(\sum X_i^2) - (\sum X_i)^2/n]\}$ and $b_0 = \bar{Y} - b_1 \bar{X}$
Coefficient of determination R^2 =0.079	So ,we have the values b_1=0.0004 ; b_0=13.287 and linear

Variance	*regression equation has view of* $Y_c=13.287-0.0004X$
$S_{y/x}=0.08$	*or* $V_k=13.287-0.0004N$

In Figure 1.5 is shown the forecasting ship speed with linear regression equation view of $V_k=13.287-0.0004N$ *(4) and determined the statistical parameters of this functional dependence:*

1.Pearson's product-moment correlation coefficient is equal

$$r= \{[\ \Sigma(X_i-\overline{X})(Y_i-\overline{Y}) \ ]\}/ \ \{[\Sigma(X_i-\overline{X})^2]^{0.50} \cdot [\ \Sigma(Y_i-\overline{Y})^2 \ ]^{0..50} \qquad where \ r=0.28$$

2.Coefficient of determination determined by formula

$$R^2 = \{ \ \Sigma XY - (\Sigma X)(\Sigma Y)/n \ \} \ / \{ [\ (\ \Sigma X)^2 -(\Sigma X)^2 /n \ ] \cdot [\ (\Sigma Y)^2 - (\Sigma Y)^2 /n \ ] \} ,where$$

$$R^2=0.079$$

3. Sum of squared deviations is equal $\quad S_{y/x} = [\ \Sigma(Y - \overline{Y})^2 \ ] / (n-1) \ ; \ S_{y/x}=0.08$

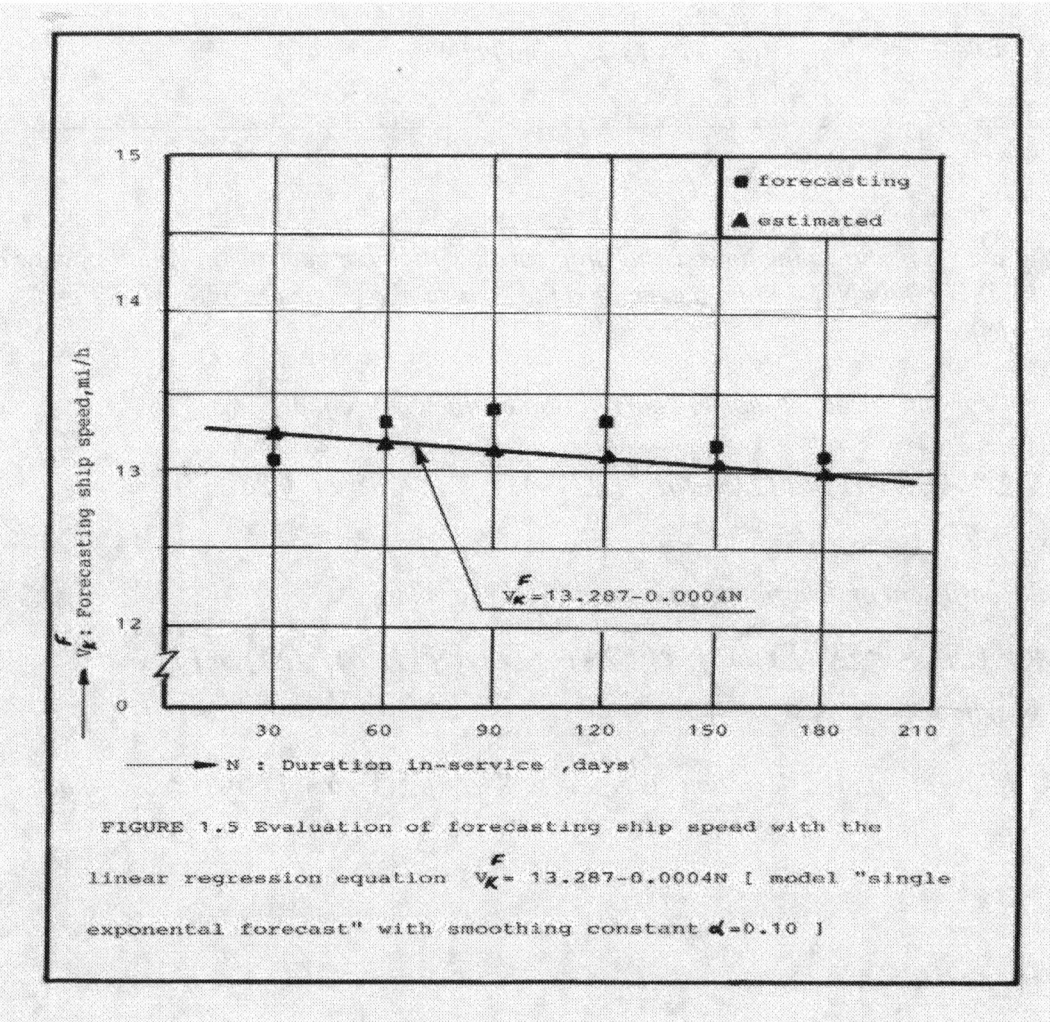

FIGURE 1.5 Evaluation of forecasting ship speed with the linear regression equation $V_K^f = 13.287 - 0.0004N$ [model "single exponental forecast" with smoothing constant $\alpha = 0.10$]

In Table 11 the calculations are shown for nonlinear regression model of view $\hat{Y}_c = a + bX + cX^2$ in question of evaluation of forecasting ship speed for the model "**single exponential forecast**" with smoothing constant $\alpha = 0.10$.

And besides is given the statistical parameters of regression equation for this functional dependence $\hat{Y}_c = 13.065 + 0.0059X - 0.000026X^2$.

Table 11 The calculations for regression equation $Y_c = 13.065 + 0.0059X - 0.000026X^2$ of forecasting ship speed for the model "single exponential forecast" with smoothing constant $\alpha = 0.10$.

Y	X	X^2	X^3	X^4	XY	X^2Y	$\hat{Y}_c$	$(\hat{Y}_c)^2$	$(Y-Y)^{-2}$	$(Y-\hat{Y}_c)^2$	Y^2
13.192	30	900	27000	$81x10^4$	395.76	11872.8	13.219	174.742	0.019	0.000	174.029

100

13.354	60	3600	216000	1296×10^4	801.24	48074.4	13.325	177.556	0.001	0.001	178.329
13.431	90	8100	729000	6561×10^4	1208.8	108791	13.385	179.158	0.010	0.002	180.392
13.365	120	14400	1728000	20736×10^4	1603.8	192456	13.399	179.533	0.001	0.012	178.623
13.336	150	22500	3375000	50625×10^4	2000.4	300060	13.365	178.623	0.000	0.001	177.849
13.30	180	32400	5832000	104976×10^4	2394	430920	13.285	176.491	0.001	0.000	176.89
Total: 79.978	630	81900	12×10^6	184275×10^4	8403.9	1092174		1066.103	0.032	0.017	1066.11

Mean : $\bar{Y}=13.329$; $\bar{X}=105$	Three normal equations for determination of parameters for the second degree curve view of $\hat{Y}_c = a + bX + cX^2$: 1. $\sum Y = na + b\sum X + c\sum X^2$ 2. $\sum XY = a\sum X + b\sum X^2 + c\sum X^3$ 3. $\sum X^2 Y = a\sum X^2 + b\sum X^3 + c\sum X^4$ $79.978 = 6a + 630b + 81900c$ $8403.99 = 630a + 81900b + 11907000c$ $1092174.30 = 81900a + 1190700b + 1842750000c$ Solving these three equations, we determine the coefficients $a = 13.065$; $b = 0.0059$; $c = -0.000026$. So, the regression equation has view : $$\hat{Y}_c = 13.065 + 0.0059X - 0.000026X^2$$
Standard error $\sigma = 0.039$	
Coefficient of determination $R^2 = 0.894$	
Correlation coefficient $r = 0.946$	

Statistical parameters of regression equations $V_k{}^F=13.065+0.0059N-0.000026N^2$:

1. *Standard error is equal* $\sigma = (\sum Y^2 - \sum Y_c^{\wedge 2})/n,$ *where* $\sigma=0.039$

2. *Coefficient of determination* $R^2 = (\sum Y_c^{\wedge 2} - \overline{Y}\sum Y)/(\sum Y^2 - \overline{Y}\sum Y),$ *where* $R=0.894$

3. *Correlation coefficient* $r=(R^2)^{0.50}$ *where* $r=0.946.$

   Analysis of forecasting ship speed with the nonlinear regression equation

$V_c{}^F=13.065+0.0059N-0.000026N^2$ **[model "single exponential forecast" with smoothing constant** $\alpha=0.10$ **]** *is shown in Figure 1.6.*

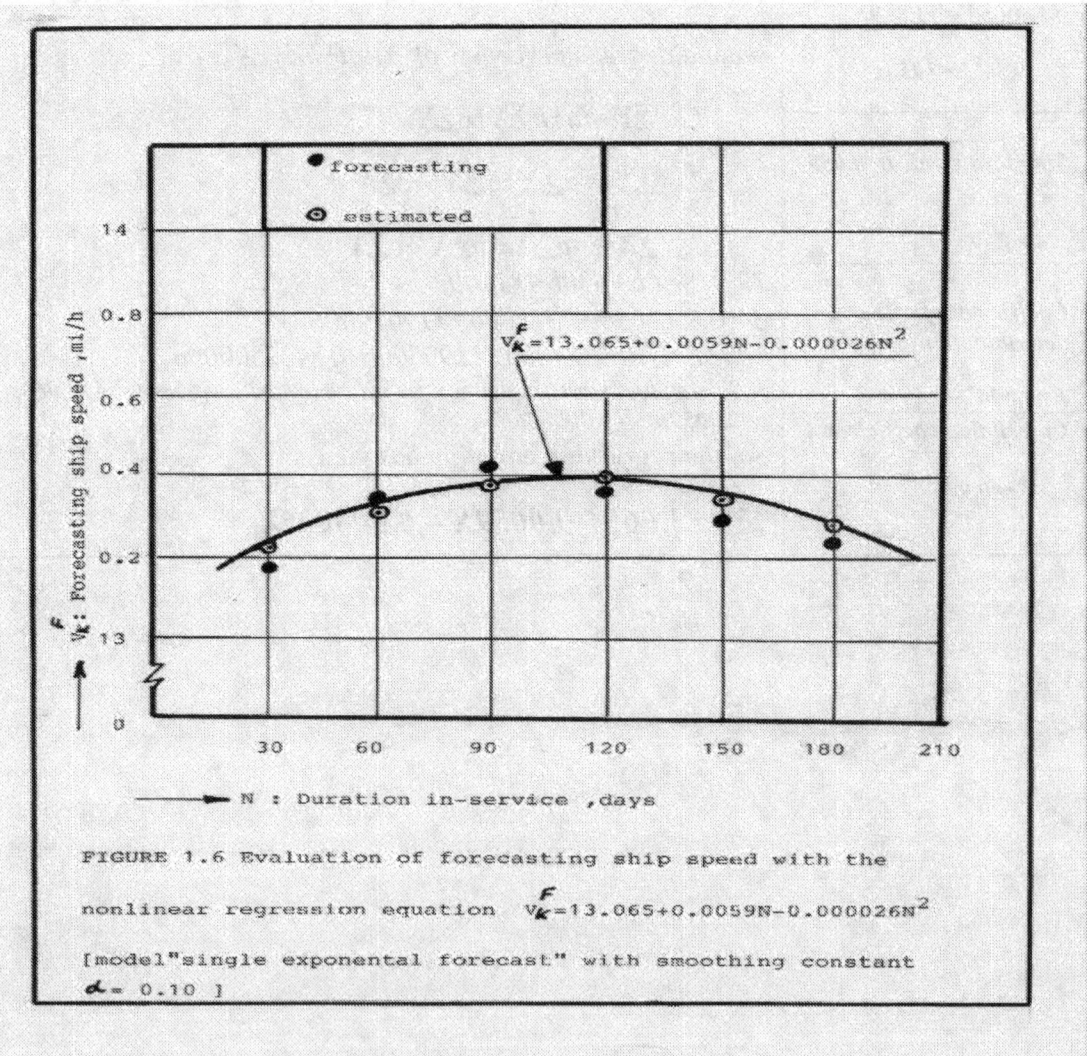

FIGURE 1.6 Evaluation of forecasting ship speed with the nonlinear regression equation $V_k{}^f=13.065+0.0059N-0.000026N^2$ [model"single exponental forecast" with smoothing constant $\alpha=0.10$]

From Figure 1.6 we see that at this case the forecasting ship speed considerably decreases with increasing of duration in-service and this fact was confirmed in above – named analysis.

In Table 12 shown the general parameters of testing and evaluation the forecasting ship speed with using **method of " single exponential forecast" with smoothing constant α=0.10.**

Table 12 Evaluation and comparison of method "single exponential forecast" with smoothing constant α=0.10 for the ship speed.

Linear model (straight line)	Nonlinear model (Second degree curve)	Nonlinear model (Exponential curve)	Results and conclusions
Regression equation $V_k = 13.287 - 0.0004N$	**Regression equation** $V_k = 13.065 + 0.0059N - 0.00002626N^2$	**Regression equation** $V_k = 13.40N^{-0.012}$	***Results and conclusions***
1.Mean: $\overline{Y} = 13.329$ $\overline{X} = 105$	1. Mean : $\overline{Y} = 13.329$ $X = 105$	1.Mean: $\overline{Y} = 13.329$ $X = 105$	Compare and analyze the above-named three regression equations: *Linear model (straight line)* $V_k = 13.287 - 0.0004N$
2.Coefficient of correlation $r = 0.28$	2.Coefficient of correlation $r = 0.946$	2.Coefficient of correlation $r = 1.00$	
3.Coefficient of determination $R^2 = 0.079$	3.Coefficient of determination $R^2 = 0.894$	3.Coefficient of determination $R^2 = 1.00$	*Nonlinear model (second degree curve)* $V_k = 13.65 + 0.0059N - 0.000026N^2$
4. Variance $S_{y/x} = 0.08$	4.Standard error $\sigma = 0.039$	4. Standard error $\sigma = 1.125$	*Nonlinear model (exponential curve)* $V_k = 13.40N^{-0.012}$
5.Mean absolute deviation $\min MAD = \sum \lvert Y - \hat{Y_c} \rvert / n$ $\min MAD = 0.508/6 = 0.085$	5.Mean absolute deviation $\min MAD = \sum \lvert Y - \hat{Y_c} \rvert / n$ $\min MAD = 0.054/6 = 0.01$	5.Mean absolute deviation $\min MAD = \sum \lvert Y - \hat{Y_c} \rvert / n$ $\min MAD = 0.012/6 = 0.002$	we use the main criteria for determining the best forecasting model and

6. Mean square and error $$\min MSE=\sum(Y-Y_c)^2/n$$ $$\min MSE =0.082/6=0.014$$	6. Mean square and error $$\min MSE=\sum(Y-Y_c)^2/n$$ $$\min MSE=0.017/6=0.028$$	6. Mean square and error $$\min MSE=\sum(Y-Y_c)^2/n$$ $$\min MSE=0.037/6=0.006$$	regression equation in view such parameters as **MSE and MAD which permit to forecast the ship speed [model "single exponential forecast" with smoothing constant $\alpha=0.10$] and make to correct choice.** *So, the forecasting ship speed submits to regression* $V_k^F=13.40N^{-0.012}$

In Table 13 is shown the calculations of nonlinear regression model $\hat{Y}_c=a+bX+cX^2$ regarding of evaluation of forecasting ship speed for the model "single exponential forecast" with smoothing constant $\alpha=0.90$. And besides is given the statistical parameters

of regression equation view of $\hat{Y}_c=15.667-0.0237X +0.000044X^2$ (6).

Table 13 Evaluation of single exponential forecasting ship speed V_k^F for the nonlinear equation $\hat{Y}_c=15.667-0.0237X +0.000044X^2$ with smoothing constant $\alpha=0.90$

Y	X	X^2	X^3	X^4	XY	X^2Y	Y^2	$\hat{Y}_c$	$\hat{Y}_c^2$	$(Y-\bar{Y})^2$	$(Y-Y_c)^2$
14.724	30	900	27×10^3	81×10^4	441.7	13251.6	216.8	14.9	224.9	0.893	0.074
14.805	60	3600	216×10^3	1296×10^4	888.3	53298	219.2	14.4	207.4	1.053	0.162
14.188	90	8100	729×10^3	6561×10^4	1276.9	114922.8	201.3	13.9	192.9	0.167	0.089
12.916	120	14400	1728×10^3	20736×10^4	1549.9	185990.4	166.8	13.5	181.1	0.745	0.293

13.057	150	22500	3375×10^3	50625×10^4	1958.6	293782.5	170.5	13.1	171.7	0.521	0.002
12.982	180	32400	5832×10^3	104976×10^4	2336.8	420616.8	168.5	12.8	164.5	0.635	0.024
Total: 82.672	630	81900	119007000	$184275 \cdot 10^4$	8452.2	1081862.1	1143.1		1142	4.014	0.644

Mean: $\bar{Y}=13.779$; $\bar{X}=105$	**Three normal equations** *for determination of parameters of* **a,b** *and*
	c for the second degree curve view of $\hat{Y}_c = a + bX + cX^2$:
Standard error $\sigma = 0.311$	$\sum Y = na + b\sum X + c\sum X^2$
Coefficient of determination $R^2 = 0.854$	$\sum XY = a\sum X + b\sum X^2 + c\sum X^3$ $\sum X^2 Y = a\sum X^2 + b\sum X^3 + c\sum X^4$
	$82.672 = 6a + 630b + 81900c$ $8452.17 = 630a + 81900b + 11907000c$
Coefficient of correlation $r = 0.924$	$1081862.1 = 81900a + 11907000b + 1842750000c$ *Solving these three normal equations we can determine the coefficients* **a=15.667; b=-0.0237;c=0.000044.** *So , the regression equation has view of* $\hat{Y}_c = 15.667 - 0.0237X + 0.000044X^2$

Statistical parameters of regression equation $V_k^F = 15.667 - 0.0237N + 0.000044N^2$ for the forecasting ship speed

1. *Standard error is equal* $\sigma^2 = (\sum Y^2 - \sum \hat{Y}^2_c)/n$ *,where* $\sigma = 0.311$

2. *Coefficient of determination* $R^2 = (\sum \hat{Y}_c^2 - \bar{Y}\sum Y)/(\sum Y^2 - \bar{Y}\sum Y)$ *,where* $R^2 = 0.854$

3. *Coefficient of correlation* $r = (R^2)^{0.50}$ *and* **r=0.924.**

Analysis of this forecasting ship speed with the nonlinear regression equation

$V_k^F = 15.667 - 0.0237N + 0.000044N^2$ **[model "single exponential forecast" with smoothing constant α=0.90]** *is shown in Figure 1.7.*

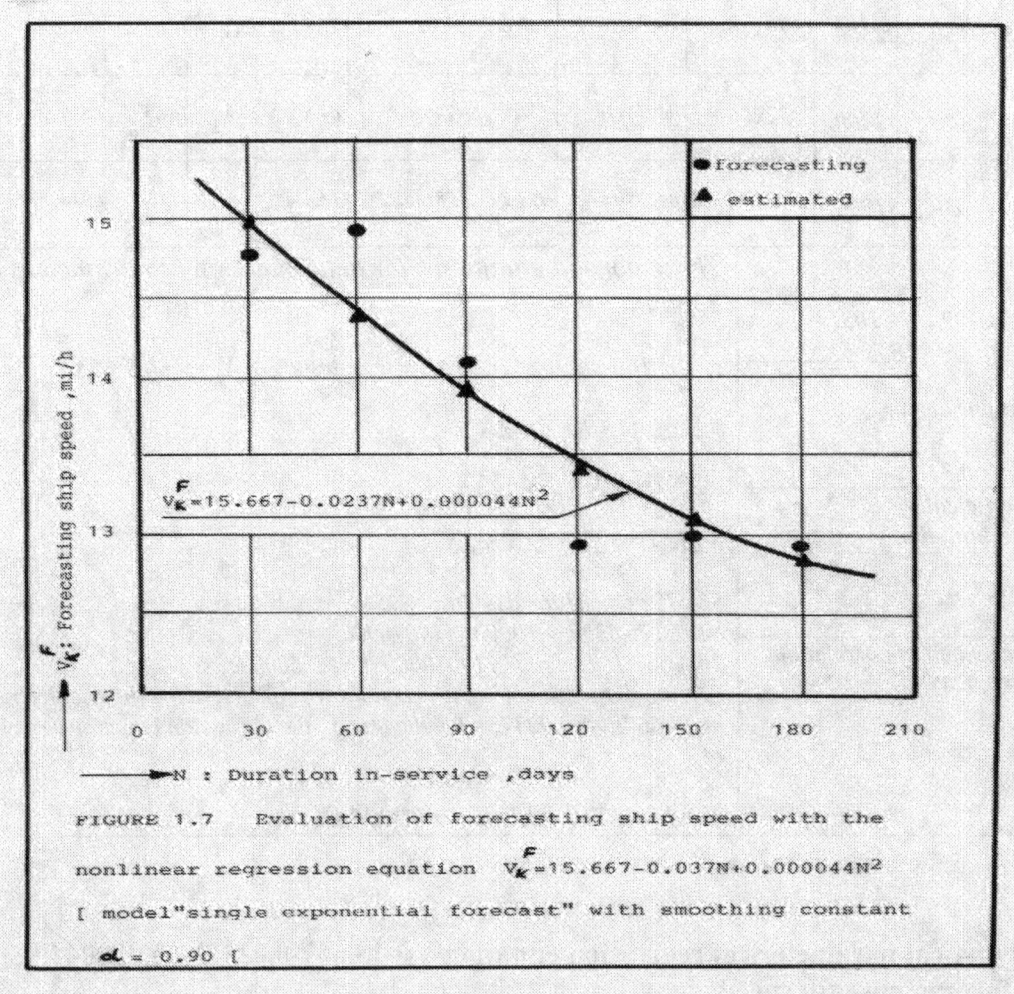

FIGURE 1.7   Evaluation of forecasting ship speed with the nonlinear regression equation $V_K^F = 15.667 - 0.037N + 0.000044N^2$ [model"single exponential forecast" with smoothing constant $\alpha = 0.90$ [

From Figure 1.7 we see that at this case the forecasting ship speed considerably decreases with increasing of duration in-service.

In Table 14 is given the calculations for determining of regression equation view of

$$\hat{Y}_c = 19.77X^{-0.081}$$ (7). This nonlinear regression model gives the possibility to evaluate the single exponential forecasting ship speed with smoothing constant $\alpha = 0.90$.

And besides this regression equation gives the possibility to compare the next forecasting models and make choice for the correct regression equation in evaluation of forecasting ship speed.

Table 14 Evaluation of single exponential forecasting ship speed for the nonlinear regression equation $\hat{Y}_c = 19.77X^{-0.081}$ with smoothing constant $\alpha = 0.90$

Y	X	logX	logY	(logX)(logY)	$(logX)^2$	$(logY)^2$	$\hat{Y}_c$	$log\hat{Y}_c$	$(log\hat{Y}_c)^2$	$(Y-\hat{Y}_c)^2$

14.724	30	1.477	1.168	1.725	2.182	1.364	15.01	1.18	1.384	0.079
14.805	60	1.778	1.170	2.080	3.162	1.369	14.20	1.15	1.330	0.372
14.188	90	1.954	1.152	2.251	3.819	1.327	13.74	1.14	1.295	0.201
12.916	120	2.079	1.111	2.309	4.323	1.235	13.42	1.13	1.272	0.257
13.057	150	2.176	1.116	2.428	4.735	1.245	13.17	1.12	1.253	0.012
12.982	180	2.255	1.113	2.509	5.086	1.239	12.99	1.11	1.240	0.000
Total: 82.672	630	11.719	6.83	13.302	23.31	7.779		6.83	7.771	0.921

Mean: $\bar{Y}=13.779$; $\bar{X}=105$	The two normal equations for calculations of coefficients **b** and **loga** : $\sum logY = nloga + b\sum logX$
Standard error $\sigma=1.138$	$\sum(logXlogY)=loga\sum logX+b\sum(logX)^2$ $6.83=6loga+b(11.719)$
Coefficient of determination $R^2=1.0$	$13.302=log(11.719)+b(23.307)$ Solving these two normal equations ,we have $b=-0.081$; $loga=1.296$ and then the regression equation has view of
Correlation coefficient $r=1.0$	$logYc=loga+blogX=1.296-0.081logX$ $\hat{Y}c=19.77X^{-0.081}$ or $Vk=19.77N^{-0.081}$

The statistical parameters of regression equation $Vk^F=19.77N^{-0.081}$:

1. Standard error is equal $\sigma^2 = \sum(log\hat{Y}_c)^2 / n$, where $\sigma=1.138$

2. Coefficient of determination $R^2 = \sum(log\hat{Y}_c)^2 / \sum(logY)^2$,where $R^2=0.999\sim1.00$

3. Coefficient of correlation $r= (R^2)^{0.50}$ and $r=1.00$

Analysis of this forecasting ship speed with the nonlinear regression equation

$Vk=19.77N^{F-0.081}$ [model "single exponential forecast" with smoothing constant $\alpha=0.90$]

is shown in Figure 1.8.

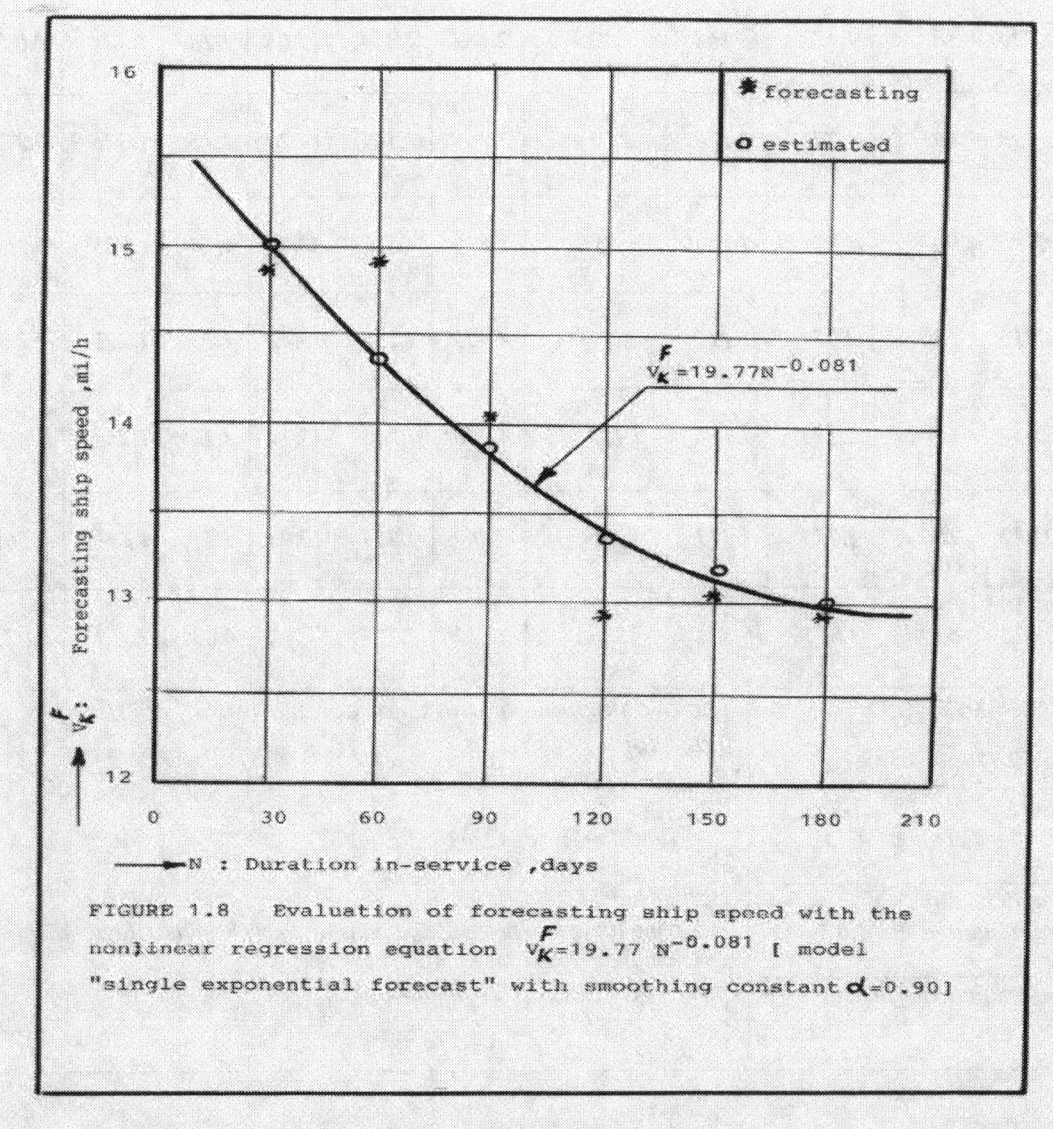

FIGURE 1.8   Evaluation of forecasting ship speed with the nonlinear regression equation $V_K^F = 19.77 \, N^{-0.081}$ [model "single exponential forecast" with smoothing constant $\alpha = 0.90$]

And besides we see that the regression model from Figure 1.8 has rather the good statistical parameters and also the same dependence was discovered in the previous regression models: **with increasing of duration in-service (N) ,the forecasting ship speed (V_k) considerably decreases.**

In Table 15 is given the data and calculation for determining of the regression equation view of $Y_c = 15.374 - 0.015X$ (8).

Table 15 Evaluation of single exponential forecasting ship speed for the linear regression equation $\hat{Y_c} = 15.374 - 0.015X$ with smoothing constant $\alpha = 0.90$

Y	X	X^2	Y^2	XY	$\hat{Y_c}$	$(Y-\bar{Y})^2$	$(Y-\hat{Y_c})^2$	$(X-\bar{X})$	$(X-\bar{X})^2$	$(X-\bar{X})(Y-\bar{Y})$

14.724	30	900	216.80	441.72	14.924	0.856	0.040	-75	5625	-69.375
14.805	60	3600	219.19	888.30	14.474	1.012	0.109	-45	2025	-45.27
14.188	90	8100	201.29	1276.9	14.024	0.151	0.027	-15	225	-5.835
12.916	120	14400	166.82	1549.9	13.574	0.779	0.433	15	225	-13.245
13.057	150	22500	170.49	1958.6	13.124	0.551	0.004	45	2025	-33.39
12.982	180	32400	168.53	2336.8	12.674	0.667	0.095	75	5625	-61.275
Total: 82.672	630	81900	1143.1	8452.2		4.016	0.708	0	15750	-228.39

Mean: $\bar{Y}=13.799; \bar{X}=105$	**Formula for determination of linear regression model view of** $\hat{Y_c}=b_0+b_1X$ **and the coefficients** b_0 **and** b_1:
Coefficient of correlation $r=-0.903$	$b_1= [\sum XY -(\sum X)(\sum Y)/n] / [\sum X^2 - (\sum X)^2 /n]$
Coefficient of determination $R^2=0.825$	$b_0=\bar{Y} -b_1\bar{X}$
Variance $S_{y/x}=0.896$	So, we have $b_1=-0.015$;$b_0=15.374$ and linear regression equation has view of $\hat{Y_c}=15.374-0.015X$ or $V_k=15.374-0.015N^F$

Evaluation of single exponential forecasting ship speed for the linear model with smoothing constant $\alpha=0.90$ *gave the possibility to calculate the statistical parameters of this function* $V_k=15.374-0.015N$:

1.*Pearson's product-moment correlation coefficient is equal:*

$$r= [\sum(X_i-\bar{X}) (Y_i-\bar{Y})] / \{ [\sum(X_i -\bar{X})^2] \cdot [\sum(Y_i-\bar{Y})^2]^{0.50} \} ,where\ r=-0.903$$

2. *Coefficient of determination* $R^2 = |\sum XY - (\sum X) (\sum Y)/n | / |\sum X^2 -(\sum X)^2/n | |\sum Y^2 - (\sum Y)^2/n |$

where $R^2=0.825.$

3. *Sum of squared deviation is equal* $S_{y/x}=\sum(Y_i-\bar{Y})^2 /n-1,$*where* $S_{y/x}=0.896$

Analysis of this forecasting ship speed with the linear regression equation ^{F}

$V_k=15.374-0.015N$ [model "single exponential forecast" with smoothing constant $\alpha=0.90$] *is shown in Figure 1.9.*

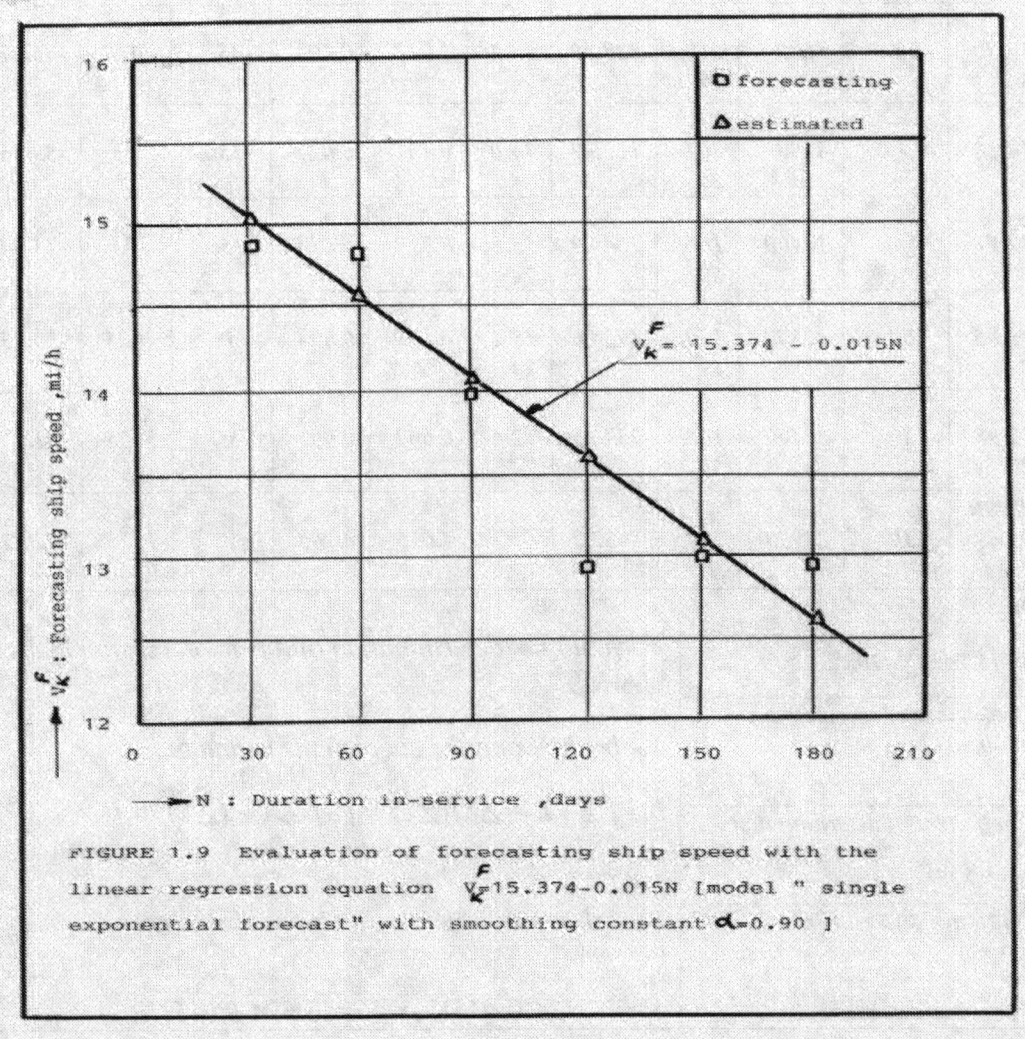

FIGURE 1.9 Evaluation of forecasting ship speed with the linear regression equation $V_k^F=15.374-0.015N$ [model " single exponential forecast" with smoothing constant $\alpha=0.90$]

  As we see from Figure 1.9 that the forecasting ship speed considerably decreases with increasing of duration in-service.
  So, after of evaluation and testing of three above-named forecasting regression equations in accordance with the single exponential forecast model at $\alpha=0.90$, we have the following data which are given in Table 16.

Table 16 Evaluation and comparison of method "single exponential forecast" with smoothing constant α=0.90 for the forecasting ship speed

Linear model (Straight line)	Nonlinear model (Second degree curve)	Nonlinear model (Exponential curve)	Result and conclusions
Regression equation F $V_k=15.374-0.015N$	Regression equation F $V_k=15.667-0.0237N+$ $+0.000044N^2$	Regression equation F $V_k=19.77N^{-0.081}$	Comparing and testing three regression equations view of: *Linear model F $V_k=15.374-0.015N$ *Nonlinear model F $V_k=15.667-0.0237N+$ $+0.000044N^2$ *Nonlinear model (Exponential curve) F $V_k=19.77N^{-0.081}$ We use the main criteria for determining the best forecasting model and regression equation in view of such parameters as **min MAD and MSE** which permit to forecast the ship speed [model single exponential forecast" with smoothing constant α=0.90. So, the regression equation is F $V_k=15.667-0.0237N+$ $+0.000044N^2$ for the forecasting ship speed
1.Mean: $\bar{Y}=13.799; \bar{X}=105$	1.Mean: $\bar{Y}=13.799; \bar{X}=105$	1.Mean: $\bar{Y}=13.779; \bar{X}=105$	
2. Coefficient of correlation r= -0.903	2.Coefficient of correlation r= 0.924	2.Coefficient of correlation r=1.00	
3.Coefficient of determination $R^2=0.825$	3.Coefficient of determination $R^2=0.854$	3.Coefficient of determination $R^2=1.00$	
4.Variance $S_{y/x}=0.896$	4.Standard error $\sigma=0.311$	4.Standard error $\sigma=1.138$	
5.Mean absolute deviation $min\ MAD=\sum|Y-\hat{Y_c}|/n$ $min\ MAD=0.122/6=0.020$	5.Mean absolute deviation $min\ MAD=\sum|Y-\hat{Y_c}|/n$ $min\ MAD=0.003/6=0.000$	5. Mean absolute deviation $min\ MAD=\sum|Y-\hat{Y_c}|/n$ $min\ MAD=0.153/6=0.026$	
6. Mean square error $min\ MSE=\sum(Y-\hat{Y_c})^2/n$ $min\ MSE=0.708/6=0.118$	6.Mean square error $min\ MSE=\sum(Y-\hat{Y_c})^2/n$ $min\ MSE=0.644/6=0.107$	6. Mean square error $min\ MSE=\sum(Y-\hat{Y_c})^2/n$ $min\ MSE=0.921/6=0.154$	

Analyzing the data which are given in Table 6 [**Evaluation and comparison of method "Moving average forecast "**], *Table 12* [**Evaluation and comparison of method" single exponential forecast " with smoothing constant α=0.10**] *for* *the forecasting ship speed and also the Table 16* [**Evaluation and comparison of method "single exponential forecast" with smoothing constant α=0.90**] *for the forecasting ship speed , we have the following data which are shown in Table 17.*

Table 17 The data for finally testing and comparing of three forecasting models for the best choice of regression equation for determining the forecasting ship speed in the tropics.

The forecasting model ' Moving average forecasts" and its regression equation (from Table 6)	The forecasting model "Single exponential forecasts with the smoothing constant α=0.10 " (Table 12)	The forecasting model " Single exponential forecasts with the smoothing constant α=0.90 (from Table 16)	Result and conclusions
Linear model (straight line) and regression equation F $Vk=16.831-0.019N$ with the data: $-F$ a. **Mean** $Vk=13.696;$ $N = 165$ b. **Coefficient of correlation** $r=0.786$ c. **Coefficient of determination** $R^2=0.993$ d. **Variance** $Sv/n=0.73$ e. **Mean absolute deviation** $\min MAD=\sum(Y-Yc)/n$ $\min MAD=0$ f. **Mean square error** $\min MSE=\sum(Y-Yc)^2/n$ $\min MSE=0.003$	**2.Nonlinear model**(exponential curve) and regression equation F $Vk=13.40N^{-0.012}$ with the data: $-F$ a. **Mean** $Vk=13.329;$ $N=105$ b. **Coefficient of correlation** $r=1.00$ c. **Coefficient of determination** $R^2=1.00$ d. **Standard error** $\sigma=1.125$ e. **Mean absolute deviation** $\min MAD=\sum(Y-Yc)/n$ $\min MAD=0.002$ f. **Mean square error** $\min MSE=\sum(Y-Yc)^2/n$ $\min MSE=0.006$	**3.Nonlinear model** (second degree curve) and regression equation F $Vk=15.667-0.0237N+$ $+0.000044N^2$ with the data: $-F$ a. **Mean** $Vk=13.799;$ $N=105$ b. **Coefficient of correlation** $r=0.924$ c. **Coefficient of determination** $R^2=0.924$ d. **Standard error** $\sigma=0.311$ e. **Mean absolute deviation** $\min MAD=\sum(Y-Yc)/n$ $\min MAD=0$ f. **Mean square error** $\min MSE=\sum(Y-Yc)^2/n$ $\min MSE =0.107$	So, analyzing and comparing three suggested forecasting models with their regression equations regarding of such important forecasting parameters as : • Mean absolute deviation (MAD); • Mean square error (MSE) and • Statistical parameters (advantageously R^2 and r) We can evaluate and make the best choice for the forecasting ship speed **as the forecast model "single exponential forecasts with the smoothing constant α=0.10,**having the nonlinear model with regression equation $Yc=13.40X^{-0.012}$ or $Vk=13.40N^{-0.012}$ for these conditions.

6.3 Interaction of total resistance and ship speed in the navigation process.

It is known that effective horsepower (e.h.p) of main engine could be determined by the formula **(Rawson ,1968)** **e.h.p=$R_T ·V$, w (9)** *where ,*

R_T =resistance of the naked hull, Newtons;

V= ship speed ,m/s

Analyzing this equation (9) we see that resistance ship is R_T= e.h.p/ V (9a).

So , the functional model of resistance ship in the process of its navigation possibility to express by the functional dependence in view of R_T= φ (V)   (9b) .

Interaction of total resistance and ship speed in the process of its navigation is shown in Figure 1.10.

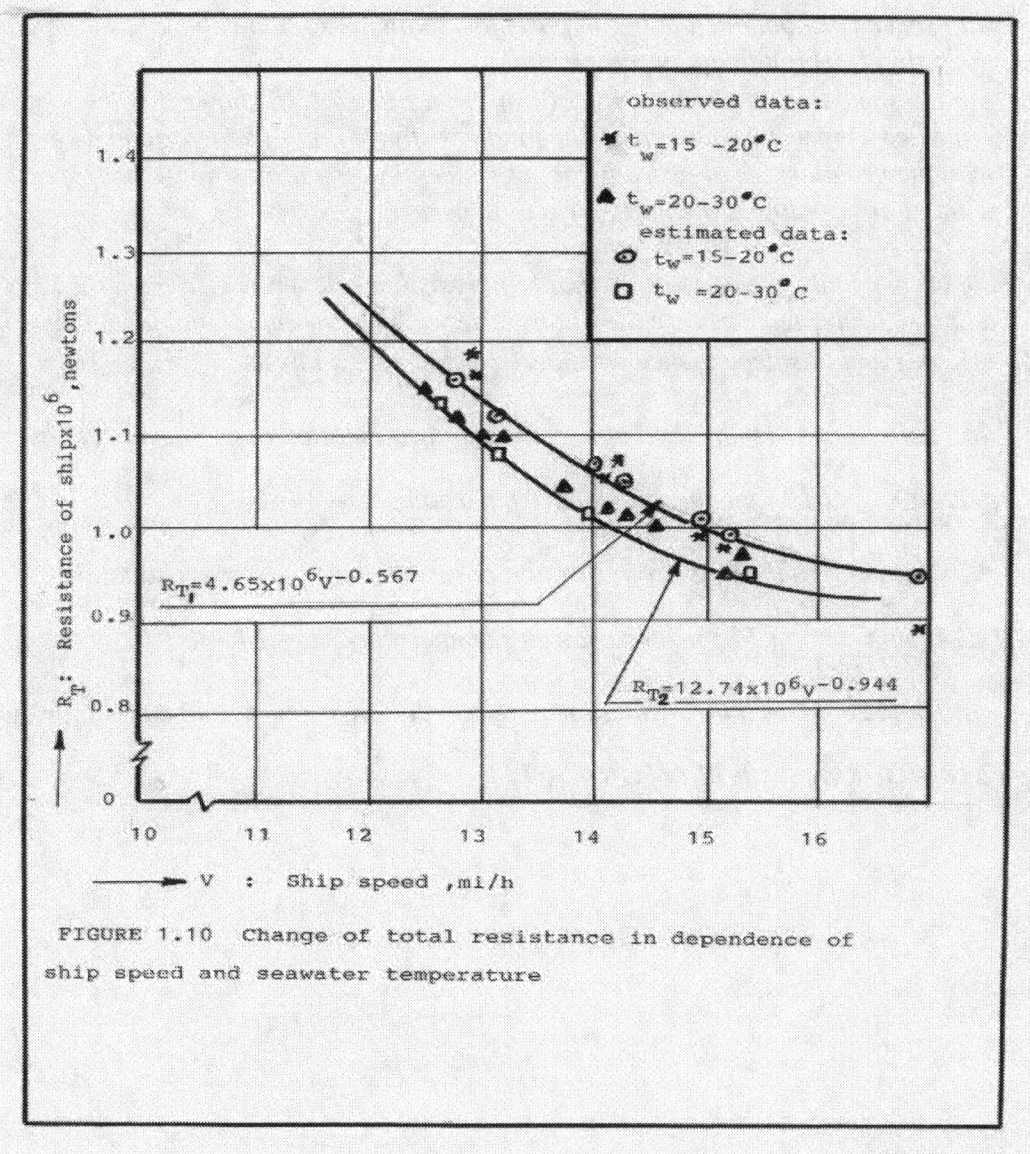

FIGURE 1.10 Change of total resistance in dependence of ship speed and seawater temperature

From Figure 1.10 we see that the total resistance of ship in dependence of ship speed has nonlinear regression models:

1. For the conditions of running ship at the seawater temperature 20 $\leq t_w \leq$ 14   the

regression equation has view of $R_{T,1}=4.65x10\,V$;

2. *And for the conditions of running ship at the seawater temperature* $30 \leq t_w \leq 20$ *the regression equation has view of* $R_{T,2}=12.74x10^{6}\,V^{-0.944}$.

- *And besides, the analysis of Figure 1.10 shows that in both cases the total resistance of ship decreases with increasing of shi speed. This mean that with increasing of ship speed (without of stopping in ports for a long time), the value of fouling on the body ship is insignificant at this period;*
- *But the same time, the navigation of ship from Figure 1.10 shows that the high seawater temperature decreases the total resistance and this factor can be explained with the peculiarities of hydrodynamics processes of motion ship in the tropics without of stopping in the ports for a long period.*

So, we can make the conclusion that total resistance depends also from the ship speed and seawater temperature, i.e at this case we have the functional model view of $R_{T}=\gamma(V\,,t_w\,)$ *and this dependence considered in the next analysis.*

In Table 18 is given the data observation and calculation of regression equation

$$\hat{Y}_c=4.65x10^{6}\,X^{-0.567} \quad (\,10\,)$$ *for the seawater temperature* $t_w = 14 \div 20$ *°C.*

And in Table 19 is given the data observation of regression equation

$$\hat{Y}_c= 12.74x10^{6}\,X^{-0.944} \quad (\,11\,)$$ *for the seawater temperature* $t_w= 20 \div 30$ *°C.*

In Table 20 is given the evaluation and calculation of regression equation view of

$$\hat{Y}_c= [2.181-0.0784V - 0.0011\,t_w\,]x10^{6} \quad (\,12\,).$$

Table 18 The data observation and calculation of regression equation
$\hat{Y_c}=4.65 \times 10^6 X^{-0.567}$ for the seawater temperature $t_w = 14 \div 20\ ^\circ C$

$(Y) \times 10^6$	X	logX	logY	(logX)(logY)	$(logX)^2$	$(logY)^2$	$(\hat{Y_c}) \times 10^6$	$(Y-\bar{Y})^2 \times 10^{12}$	$(Y-\hat{Y_c})^2 \times 10^{12}$	$(log\hat{Y_c})^2$
0.878	16.775	1.225	5.943	7.280	1.501	35.33	0.939	0.025	0.004	35.673
0.985	14.956	1.175	5.993	7.042	1.380	35.92	1.00	0.025	0.001	36.000
0.979	15.043	1.177	5.991	7.051	1.386	35.89	0.999	0.003	0.001	35.995
1.049	14.031	1.147	6.021	6.906	1.316	36.25	1.040	0.001	0.001	36.205
1.042	14.138	1.150	6.018	6.921	1.323	36.22	1.036	0	0.001	36.185
1.153	12.775	1.106	6.062	6.705	1.224	36.75	1.097	0.014	0.003	36.484
1.159	12.713	1.104	6.064	6.695	1.219	36.77	1.099	0.015	0.004	36.494
Total: 7.245	100.431	8.084	42.092	48.60	9.349	253.1		0.082	0.015	253.04

Mean :
$\bar{Y} = 1.035 \times 10^6$
$\bar{X} = 14.347$

Standard error
$\sigma = 0.108$

Coefficient of determination
$R^2 = 0.854$

The two normal equations for calculation of coefficients **b** and **loga** :

1. $\Sigma logY = nloga + b\Sigma logX$

2. $\Sigma (logX)(logY) = loga\Sigma logX + b \Sigma (logX)^2$

Solving these two normal equations we have :
b= -0.567 ; log a =6.667
and then the equation has view of

$log \hat{Y_c} = loga + blogX = 6.667 - 0.567 logX$; $\hat{Y_c} = 4.65 \times 10^6 X^{-0.567}$

Coefficient of correlation r=0.924	

Table 19 The data observation and calculation of regression equation view of $\hat{Y}_c = 12.74 \times 10^6 X^{-0.944}$ for the seawater temperatures $t_w = 20 \div 30\,^\circ C$.

$(Y) \times 10^6$	X	logX	logY	(logX)(logY)	$(logX)^2$	$(logY)^2$	$(\hat{Y}_c) \times 10^6$	$(Y-\overline{Y})^{-2} \times 10^{12}$	$(Y-Yc)^{\wedge 2} \times 10^{12}$	$(logYc)^{\wedge 2}$
0.968	15.225	1.182	5.985	7.074	1.398	35.820	0.97	0.014	0.000004	35.841
1.026	14.361	1.157	6.011	6.955	1.339	36.132	1.030	0.004	0.00002	36.154
1.039	14.169	1.151	6.017	6.926	1.326	36.204	1.043	0.002	0.00002	36.219
1.028	14.332	1.156	6.012	6.949	1.337	36.144	1.032	0.003	0.00002	36.164
0.948	15.543	1.192	5.977	7.125	1.419	35.725	0.956	0.019	0.00006	35.766
1.067	13.80	1.139	6.028	6.866	1.299	36.337	1.069	0.001	0.000002	36.349
1.124	13.108	1.118	6.051	6.765	1.249	36.615	1.123	0.001	0.000001	36.607
1.167	12.619	1.101	6.067	6.679	1.212	36.808	1.164	0.007	0.000009	36.796
1.116	13.20	1.121	6.048	6.779	1.256	36.578	1.116	0.001	0	36.574
1.124	13.106	1.117	6.051	6.759	1.249	36.615	1.123	0.001	0.000001	36.607
1.123	13.119	1.117	6.050	6.758	1.249	36.603	1.122	0.001	0.000001	36.602
1.124	13.10	1.117	6.051	6.759	1.248	36.615	1.123	0.001	0.000001	36.607
1.122	13.125	1.118	6.049	6.762	1.250	36.590	1.121	0.001	0.000001	36.597
1.124	13.10	1.117	6.051	6.759	1.248	36.615	1.123	0.001	0.000001	36.607
1.133	13.0	1.114	6.054	6.744	1.240	36.651	1.131	0.002	0.000004	36.644
1.145	12.867	1.109	6.059	6.719	1.231	36.711	1.142	0.003	0.000009	36.695
Total: 17.37	217.774	18.126	96.561	109.378	20.550	582.76		0.061	0.0001	582.829

Mean : $\overline{Y} = 1.086 \times 10^6 X$; $\overline{X} = 13.611$	The two normal equations for calculation of coefficients **b** and **loga :**

116

Standard error σ=0.064	
Coefficient of determination R^2 =0.998	
Coefficient of correlation r=0.999	

Table 20 Evaluation and calculation of regression equation view of $R_T = [\,2.181 - 0.0784V - 0.0011t_w\,] \times 10^6$.

(Y) $\times 10^6$	X_1	X_2	X_2^2	X_1X_2	(X_1Y) $\times 10^6$	(X_2Y) $\times 10^6$	X_1^2	(Y_c) $\times 10^6$	$(Y-\bar{Y})^2$ $\times 10^{12}$	$(Y-Y_c)^2$ $\times 10^{12}$	$(Y-\bar{Y})(Y_c-\bar{Y})$ $\times 10^{12}$
0.878	16.775	19	361	318.725	14.728	16.682	281.40	0.845	0.037	0.001	0.044
0.985	14.956	15.25	232.563	228.079	14.732	15.021	223.68	0.991	0.007	0.00003	0.007
0.979	15.043	16.75	280.563	251.970	14.727	16.398	226.29	0.984	0.008	0.00003	0.008
1.049	14.031	17.25	297.563	242.034	14.719	18.095	196.87	1.063	0.0005	0.0002	0.0002
1.042	14.138	19.75	390.063	279.226	14.732	20.579	199.88	1.052	0.0008	0.0001	0.0005
1.153	12.775	18.86	355.699	240.937	14.729	21.746	163.20	1.159	0.007	0.00004	0.007
1.159	12.713	19.50	380.25	247.904	14.734	22.601	161.62	1.163	0.008	0.00002	0.008
0.968	15.225	23.75	564.06	361.594	14.738	22.990	231.80	0.961	0.011	0.00005	0.011
1.026	14.361	24.00	576.00	344.664	14.734	24.624	206.24	1.029	0.002	0.002	0.002
1.039	14.169	24.75	612.563	350.682	14.722	25.715	200.76	1.043	0.001	0.00002	0.0009
1.028	14.332	27.50	756.25	394.13	14.733	28.270	205.41	1.027	0.002	0.00001	0.002
0.948	15.543	28.00	784.00	435.204	14.735	26.544	241.58	0.931	0.015	0.0003	0.017
1.067	13.80	27.75	770.063	382.95	14.725	29.609	190.44	1.068	0.0001	0.00001	0.00001
1.124	13.108	28.00	784.00	367.024	14.733	31.472	171.82	1.122	0.003	0.00001	0.003
1.167	12.619	26.50	702.25	334.404	14.726	30.926	159.24	1.163	0.009	0.00002	0.009
1.116	13.20	22.75	517.625	300.300	14.731	25.389	174.24	1.121	0.002	0.00003	0.002
1.124	13.106	22.25	495.063	291.609	14.731	25.009	171.77	1.129	0.003	0.00003	0.003
1.123	13.119	24.25	588.063	318.136	14.733	27.232	172.11	1.125	0.003	0	0.003
1.124	13.10	27.75	770.06	363.525	14.724	31.191	171.61	1.124	0.003	0	0.003
1.122	13.125	25.00	625.00	328.125	14.726	28.05	172.27	1.124	0.003	0.00001	0.003
1.124	13.10	23.25	540.56	304.575	14.724	26.133	171.61	1.128	0.003	0.00002	0.003
1.133	13.00	22.00	484.00	286.00	14.729	24.926	169.00	1.138	0.004	0.00003	0.004
1.145	12.867	21.00	441.00	270.207	14.733	24.045	165.56	1.149	0.005	0.00002	0.006
Total: 24.623	318.205	524.86	12308.30	7242.00	338.778	563.251	4428.4		0.137	0.004	0.146

Mean: $\bar{Y}=1.071\times10^6$; $\bar{X_1}=13.835$; $\bar{X_2}=22.82$	The solving three normal equations $$\sum Y = nb_0 + b_1\sum X_1 + b_2\sum X_2$$ $$\sum X_1Y = b_0\sum X_1 + b_1\sum X_1^2 + b_2\sum X_1X_2$$
Coefficient of determination $R^2=0.971$	$$\sum X_2Y = b_0\sum X_2 + b_1\sum X_1X_2 + b_2\sum X_2^2$$ we can determine the coefficients b_0, b_1 and b_2. So ,at data $$24.623\times10^6 = 23b_0 + 318.205b_1 + 524.86b_2$$
Coefficient of multiple correlation $r=0.954$	$$338.778\times10^6 = 318.205b_0 + 4428.394b_1 + 7242.004b_2$$
Standard error $S_{y/x}=0.014\times10^6$	$$563.251\times10^6 = 524.86b_0 + 7242.004b_1 + 12308.256b_2$$ where $b_0=-0.128\times10^6$; $b_1=0.072\times10^6$; $b_2=0.007\times10^6$ and regression
Regression sum of squares $SSR=0.133\times10^{12}$	equation $$\hat{Y_c} = [2.181 - 0.0784X_1 - 0.0011X_2]\times10^6$$

In Table 21 is given the calculation of statistical parameters for the regression equations $\hat{Y}_c = 4.65 \times 10^6 \, X^{-0.567}$ and $\hat{Y}_c = 12.74 \times 10^6 \, X^{-0.944}$.

Table 21 The calculation of statistical parameters

Statistical parameters	Regression model view of $\hat{Y}_c = 4.65^6 \times 10 \, X^{-0.567}$	Regression model model view of $\hat{Y}_c = 12.74^6 \times 10 \, X^{-0.944}$
Standard error $\sigma^2 = [\; \Sigma(\log Y)^2 - \Sigma(\log \hat{Y}_c)^2 \;] / n$	0.108	0.064
Coefficient of determination $R^2 = [\; \Sigma(Y-\bar{Y})^2 - \Sigma(Y-\hat{Y}_c)^2 \;] / \Sigma(Y-\bar{Y})^2$	0.854	0.998
Coefficient of correlation $r^2 = (R)^{0.50}$	0.924	0.999

Evaluation of statistical parameters of regression equation view of

$$\hat{Y}_c = (\, 2.181 - 0.0784 \, X_1 - 0.0011 \, X_2 \,) \times 10^6 \; :$$

1.Coefficient of determination is equal $R = [\Sigma(Y-\bar{Y})^2 - \Sigma(Y-\hat{Y}_c)^2] / \Sigma(Y-\bar{Y})^2$ where $R^2 = 0.971$.

2.Coefficient of multiple correlation is

$r = \Sigma(Y-\bar{Y})(\hat{Y}_c-\bar{Y}) / \{ [\Sigma(Y-\bar{Y})^2 \cdot \Sigma(\hat{Y}_c-\bar{Y})^2] \}$ *(Kleinbaum,1978)* where r = 0.954

3. Standard error $S_{y/x1,x2} = \Sigma(Y-\hat{Y}_c)^2 / (n-3)$, where $S_{y/x1,x2} = 0.014 \times 10^6$;

4. Regression sum of squares SSR is equal $SSR = (\Sigma Y-\bar{Y})^2 - (\Sigma Y-\hat{Y}_c)^2$, where $SSR = 0.133 \times 10^{12}$;

5. Analysis and calculation of F-value showed the following results:

$$F = [SSR/k] / [\, SSE / (n-k-1) \,] = \{ [(\Sigma Y-\bar{Y})^2 - \Sigma(Y-\bar{Y})^2] \times (n-k-1)\} / \{ k[\Sigma(Y-\hat{Y}_c)^2] \},$$
where k=2 ,n-k-1=20 and **F=332.50**

Referring to recommendation (**Pfaffenberger,1977)** from Table 7,Appendix A the critical F-value at the $\alpha=0.05$ $F_{0.05;2;20}=3.49$ and as $F > F_{0.05;2;20}$,we reject the hypothesis that all regression coefficients are zero and this conclusion we see in the regression equation view of

$$\hat{Y}_c = (2.181 - 0.0784X_1 - 0.0011X_2) \times 10^6 \quad \text{or} \quad R_T = [2.181 - 0.0784V - 0.0011t_w] \times 10^6.$$

Scatter plots of ship speed (V) and seawater temperature (t_w) against the total resistance (R_T) are illustrated in Figure 1.11.

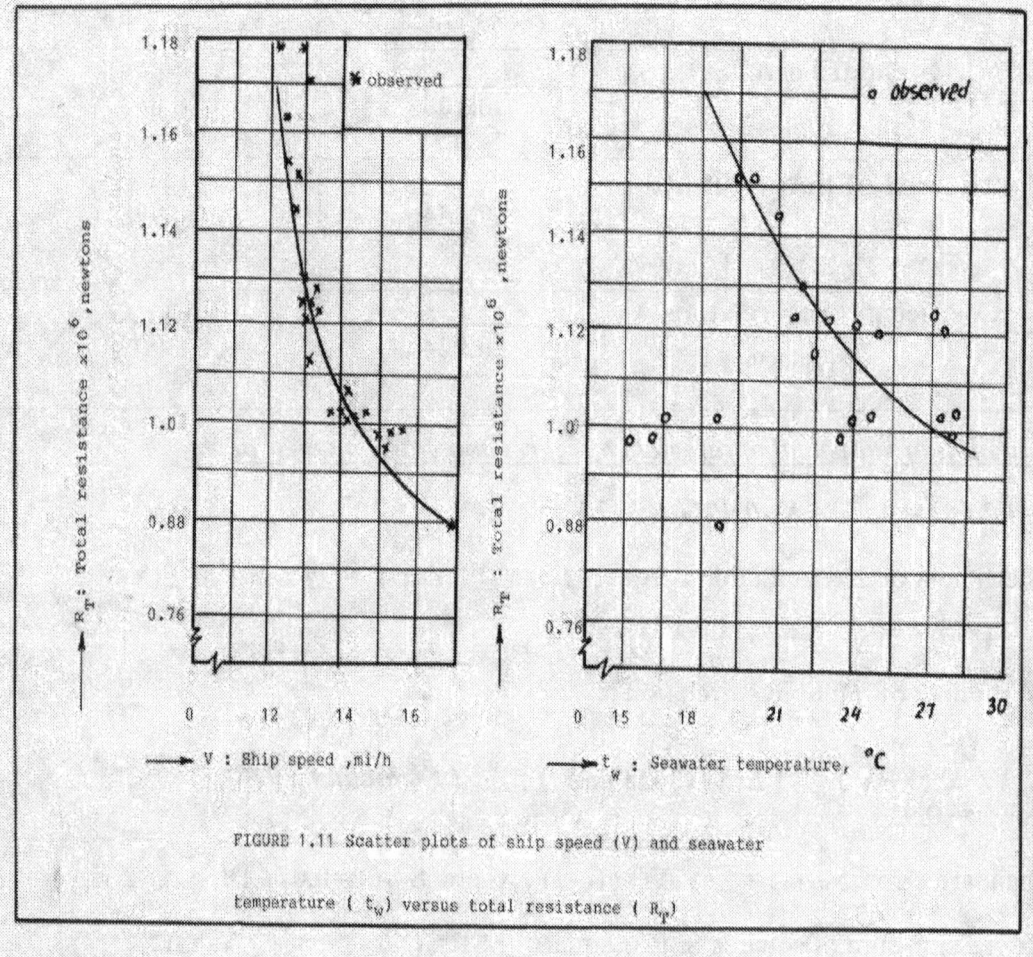

FIGURE 1.11 Scatter plots of ship speed (V) and seawater

temperature (t_w) versus total resistance (R_T)

As we see from Figure 1.11 the relationship between ship speed and total resistance has the nonlinear model and this functional dependence in detail is shown in previous Figure 1.10.

And the second independent variable, such as the seawater temperature (t_w) has also the nonlinear relationship and this regression model characterizes by the following functional dependence: the total resistance (R_T) decreases with increasing of the seawater temperature (t_w).

6.4 Prediction of frictional resistance, as the index of fouling, of body ship in the tropics.

Referring to the recommendation of author **(Rawson,1968)** ,we can calculate the principal dimensions of our cargo ship which was described in the previous papers:

- **L** =length of ship is **L=600feet (182 m);**
- **B**= width of ship is equal **B=L^n,** where n=0.675÷0.680.
  So, the value **B=77feet(23m);**
- **T**= draught , **T=33feet (10m).**

It is necessary to admit that at the beginning navigation process, the body of ship is cleaned and painted after of dock works and we can evaluate that at this case the skin frictional resistance (**f$_0$**)is equal **f$_0$=0.07.**

But after of six-month navigation period for the ship in the tropics, this skin frictional resistance considerably increases and these conclusions very well coordinate with the data of some authors, indicating on the fact that on the body ship appears the fouling.

The frictional resistance (**R$_F$**) can be calculated by formula

$$\mathbf{R_F = f \cdot S \cdot V^{1.825}} \quad \text{, Newtons} \quad (13)$$

where,

f = Fraude's skin friction constant (for L=600 feet we have f =0.070) ;
V = speed of ship , m/s ;
S = wetted surface , m^2 ; where **S = (3.4 + M^2/2.06)** **(14)**
M= length constant , M= L/U ;

U = characteristic length ,ft and **U= ($\nabla \cdot$ R)$^{1/3}$** **(15)** ,where ∇=displacement ,tonf

where [deadweight/displacement =0.67 **(16)**] and we have ∇= 18,307 (deadweight/0.67=10,984/0.67);

R = minimum residuary resistance R =55 for V / (L)$^{0.50}$ = [14.10/ (600)$^{0.50}$]=0.60
So ,we finally have U= 100ft and M= 6.0 , S=6.312.

a) Frictional resistance (R$_F$) of ship at the starting navigation period in the tropics

Accordingly with the above-named formulas we have the following data:

f$_0$ =0.070 ; S=6.312 ; V=14.10 Knots (7.25 m/s) we have the frictional resistance
equal **R$_F$ = f$\cdot$S$\cdot$V$^{1.825}$ =0.070$\cdot$6.312$\cdot$ [(14.10$\cdot$1852)/3600]$^{1.825}$ = 16.419 Newtons.**

b) Frictional resistance (R $_F$) at the ending navigation process , after of six-months running in the tropics.

Statistical analysis given in above-named papers showed that the ship speed considerably decreased, about of 30 percentage, when the ship run in the tropics .So, we can conclude that at end of navigation of ship in the tropics, the frictional resistance (R_F) considerably increases.

In view of that ship was in the six-months navigation in the tropics, we can conclude that skin friction resistance increases in a quarter of one percent per day (**Rawson,1968**).

So, we can calculate the value $f = 1/4 \cdot 0.01 \cdot N$ (**17**)
where **N**= duration in-service of ship in the tropics (N=180 days) and then f =0.450.

And then for the frictional resistance R_F , we have the following results :

$$R_F = [\, f \cdot S \cdot V^{1.825}\,] \quad (18)$$ at data f =0.450 ; S=6.312; V =11.2 Knots (5.76 m/s) and

$$R_F = 0.450 \cdot 6.312[\,(11.2 \cdot 1852)/3600\,]^{1.825} = 73.99 \text{ Newtons.}$$

The relative increasing of frictional resistance ,as the index of fouling ,in the process of six-months running ship in the tropics from the starting period is equal :

$$K = (\, R_F - R_F) / R_F \quad (19)$$ where K=0.76 or 76 %.

As we see that with increasing of duration in-service of ship in the tropics more than on six-months navigation period, its frictional resistance increased more than on 76 percentage and accordingly the ship speed decreased about of 20 ÷30 percentage.

In Table 22 is shown the evaluation of frictional resistance (index of fouling) and reduction of ship speed.

Change of the frictional resistance and reduction of ship speed in dependence of duration in-service of ship in the tropics is shown in Figure 1.12.

As we see from Figure 1.12 the frictional resistance has the linear model and this value (index of fouling) considerably increases with increasing of duration in-service of ship in the tropics.

And besides we see from Figure 1.12 that reduction ship speed (ΔV) has the nonlinear model which has the regression equation view of $\Delta V = 1.33N^{0.54}$ (**20**).

From Figure 1.12 we also see that reduction ship speed increases with increasing of duration in-service ship in the tropics.

Table 22 Evaluation of frictional resistance (index of fouling) and reduction of ship speed

Duration in-service N,days	Ship speed V			Fraude's skin frictional resistance, f	Wetted surface, S (m²)	Frictional resistance, RF,Newtons	Monthly reduction ship speed, ΓV, in percentage
	mi/h	Knots	m/s				
1	16.4	14.10	7.25	0.07	6.312	16.419	-
30	14.92	12.83	6.60	0.075	6.312	14.826	9.04
60	14.81	12.74	6.56	0.150	6.312	29.316	9.67
90	14.12	12.14	6.24	0.225	6.312	40.139	13.90
120	12.78	10.99	5.65	0.30	6.312	44.645	22.10
150	13.07	11.24	5.78	0.375	6.312	58.172	20.29
180	12.97	11.16	5.74	0.450	6.312	68.927	20.89

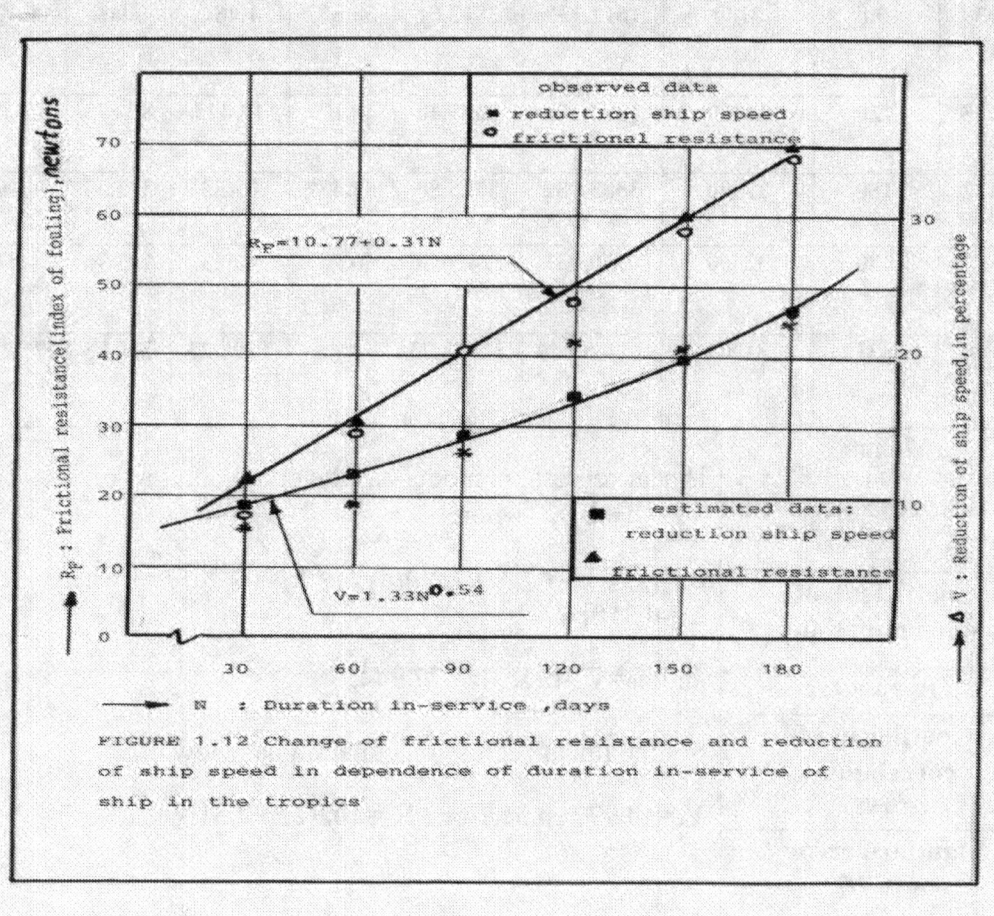

FIGURE 1.12 Change of frictional resistance and reduction of ship speed in dependence of duration in-service of ship in the tropics

Table 23 Calculation and evaluation of linear regression equation view of

R_F= 10.977+0.31N for the frictional resistance (index of fouling).

Y	X	X^2	Y^2	XY	$\hat{Y}c$	$(Y-\bar{Y})^2$	$(\hat{Y}c-\bar{Y})^2$	$(Y-Yc)^2$
16.419	1	1	269.584	16.419	11.287	506.34	763.64	26.337
14.826	30	900	219.810	444.78	20.277	580.569	347.59	29.713
29.316	60	3600	859.427	1758.96	29.577	92.256	87.31	0.068
40.139	90	8100	1611.139	3612.51	38.877	1.484	0.002	1.593
44.645	120	14400	1993.176	5357.40	48.177	12.475	85.673	12.475
58.172	150	22500	3383.982	8725.80	57.477	0.483	344.33	0.483
68.927	180	32400	4750.931	12406.86	66.777	4.623	775.96	4.623
Total: 272.444	631	81901	13088.049	32322.73		1198.23	2404.5	75.292

Mean : $\bar{Y}$ = 38.921 ; $\bar{X}$ = 90.143	Linear regression model has the equation view of: $Y = b_0+b_1X$ where $b_1=[\sum X_iY_i- (\sum X_i)(\sum Y_i)/n] / [\sum X_i^2 - (\sum X_i)^2 /n]$ where $b_1=0.310$;
Coefficient of determination R^2 =0.969	
Coefficient of correlation r= 0.985	and $b_0=\bar{Y} -b_1\bar{X}$, $b_0=27.944$. So ,the linear regression equation has view of $\hat{Y}_c=10.977+0.31X$ or $\hat{R}_F=10.977+ 0.31N$
Standard error $S_{y/x}=7.97$	

Statistical parameters of regression equation $R_F=10.917+0.31N$:

1.Coefficient of determination $R^2 = \left|\sum XY - (\sum X)(\sum Y)/n\right| / \{\left[\sum X^2 - (\sum X)^2/n\right] \cdot \left[\sum Y^2 - (\sum Y)^2/n\right]\}$

where $R^2=0.969$;

2. Coefficient of correlation

$\quad r=[\sum X_iY_i - (\sum X_i)(\sum Y_i)/n] / \{[\sum X_i^2 - (\sum X_i)^2/n]^{0.50} \cdot [\sum Y_i^2 - (\sum Y_i)^2/n]^{0.50}\}$, where

$r=0.985$

3. Standard error $S_{y/x}=(1/n-2)\{ \left|\sum Y^2 - (\sum Y)^2/n\right| - \left|\sum XY - (\sum X)(\sum Y)/n\right|^2 / \left|\sum X^2 - (\sum X)^2/n\right|$

where $S_{y/x}=7.97$

The calculation and evaluation of non-linear regression function $\Delta V= \gamma (N)$ is shown in Table 24.

Table 24 Evaluation and calculation of regression equation $\Delta V=1.33N^{0.54}$

Y	X	logX	logY	(logX)(logY)	(logX)^2	(logY)^2	$(\hat{Y}c)$	$\wedge$ logYc	$\wedge^2$ (logYc)
9.04	30	1.477	0.956	1.412	2.182	0.914	8.35	0.922	0.849
9.67	60	1.778	0.985	1.751	3.162	0.971	12.14	1.084	1.175
13.90	90	1.954	1.143	2.233	3.819	1.306	15.11	1.179	1.391
22.10	120	2.079	1.344	2.794	4.322	1.807	17.64	1.246	1.553
20.29	150	2.176	1.307	2.844	4.735	1.709	19.90	1.299	1.687
20.89	180	2.255	1.319	2.974	5.086	1.742	21.96	1.342	1.799
Total: 95.89	630	11.719	7.054	14.008	23.306	8.50			8.45

Mean:	
$\overline{Y}= 15.982$; $\overline{X}=105$	The normal two equations for determining of coefficients **b** and **loga** have the following view:
Standard error $\sigma=0$	$\sum\log Y = n\log a + b\sum\log X$ $\sum(\log X\log Y)=\log a\sum\log X + b\sum(\log X)^2$

Coefficient of determination $R^2 = 1.0$	
Coefficient of correlation r=1.00	

Statistical parameters of regression equation $\Delta V = 1.33N^{0.54}$:

1. Standard error $\sigma = [\sum(\log Y)^2 - \sum(\log Y_c)^2]/n$,where $\sigma=0$;

2.Coefficient of determination $R = \{\sum(\log Y_c)^2 - (\overline{\log Y})\sum\log Y\}/\{\sum(\log Y)^2 - (\overline{\log Y})\sum\log Y\}$

where $R^2 = 1.0$;

3.Coefficient of correlation $r = (R^2)^{0.50}$,   where r=1.0.

CHAPTER 7

Evaluation and prediction of fouling ship's hull in the tropics

7.1 Influence of fouling on the main parameters of running ship and methods of its estimation

As was indicated above, the reduction ship speed (ΔV) increases with increasing of duration in-service ship in the tropics. And necessary to admit that the parameter of fouling very difficult to estimate in process of running ship in the ***tropics and author of this paper suggests to use the indirect method of evaluation this parameter which consists in the following three steps:***

1. *At the first step,* in starting period of navigation ship in the tropics necessary to make the external examination of body ship and evaluate the value of fouling. This process usually makes in dock and statistical shows that in starting period of ship, before of its long navigation in the tropics, the body of ship usually is clean and this value we can approximately to accept equal as **δ=0.05 mm;**

    2.*At the second step* ,in navigation period periodically evaluates the value of reduction speed (ΔV) in percentage, relatively of starting ship speed (V_0) and then calculates the average fouling which has the body of ship ,taking into account **that one percent of reduction speed for the ship in tropics is equal to the statistical coefficient of fouling C=0.30÷0.50 mm** which was determined by the author with using of the statistical methods.

So , we can calculate for some period of running ship its average monthly fouling (δ_i) to the formula:

$$\delta_i = (C \cdot \Delta V) / (K \cdot 100) \qquad (1)$$

where,

        δ_i = average monthly fouling ship's hull ,mm   ;

        ΔV = monthly reduction of ship speed, in percentage;

        K= experimental coefficient which is equal K=0.01;

        C= statistical coefficient of fouling which is equal C=0.30 ÷0.50 mm.

As the example, in Table 1 is shown the second step of estimation of fouling for the six-month period of navigation ship in the tropics.

Table 1 Estimation of fouling for the six-month period of navigation ship in the tropics

Months of running ship	Duration in-service N, days	Monthly reduction ship speed ,in percentage ΔV	Average Monthly Fouling δ_i, mm	Cumulative fouling for the running period ($\sum\delta$) ,mm
August	30	9.04	4.70	4.70
September	60	9.67	5.02	9.72
October	90	13.90	7.23	16.95
November	120	22.10	11.49	28.44
December	150	20.29	10.55	38.99
January	180	20.89	10.86	49.85

3. *At the third step,* in ending period of navigation ship in the tropics, necessary again to make the external examination of body ship and determine actual value of fouling ship's hull in the dock.

- For the cargo ship, parameters which was described in above-named papers, at the ending running period of ship, the actual value of fouling after of six-month navigation period was equal $\sum\delta$ =45 mm. In Table 1 is shown that calculated ending fouling value is equal $\sum\delta$=49.85 mm and the relative error of calculation and evaluation is equal about of 9.70 percentage.

- Analysis of navigation ship in the tropics shows that the value of fouling is the complex parameter and submits to the multiple regression analysis.

On the basis of these conclusions, we can accept into account that the fouling functionally is joined with such parameters as:

duration in-service (N) ,revolution of engine (n) and ship speed (V) ,i.e we can this model to write in view of $\sum \delta = \phi (N, n, V)$ (2) .

In Table 2 is shown calculation and evaluation of this functional dependency and regression equation view of $\sum \delta$ = - 140.935+0.359N+1.349n-0.826V (3)

Table 2 Calculation and evaluation of regression equation $\sum \delta$=-140.935+0.359N+1.349n-0.826V

Y	X_1	X_2	X_3	X_1Y	X_2Y	X_3Y	X_1^2	X_1X_2	X_1X_3	X_2^2	X_2X_3	X_3^2	$\hat{Y}c$	$(Y-\bar{Y})^{-2}$	$(Yc-\bar{Y})^{-2}$	$(Y-Yc)^{\wedge 2}$	$(Y-\bar{Y})\cdot(Yc-\bar{Y})$
4.70	30	109	14.9	141	512	70.11	900	3270	447.5	11881	1625.8	222.5	4.56	403	408.6	0.02	405.8
9.72	60	104	14.8	583	1010	143.99	3600	6240	888.84	10816	1540.7	219.5	8.67	226.7	258.7	1.103	242.16
16.95	90	103	14.1	1525	1745	239.33	8100	9270	1270.8	10609	1454.4	199.4	18.7	61.23	37.39	2.924	47.85
28.44	120	101	12.8	3412	2872	363.32	14400	12120	1533	10201	1290.3	163.2	27.8	13.43	9.41	0.358	11.24
38.99	150	102	13.1	5848	3976	509.72	22500	15300	1960.9	10404	1333.5	170.9	39.7	202.1	223.35	0.533	212.44
49.85	180	101	12.9	8973	5034	646.75	32400	18180	2335.3	10201	1310.4	168.3	49.2	628.8	597.56	0.397	612.95
Total: 148.7	630	620	82.7	20484	15153	1973.2	81900	64380	8436.4	64112	8554.9	1143		1535	1535.1	5.335	1532.5

Mean:	The coefficients **b0,b1,b2** and **b3** can be determined from solving four normal equations:
$\bar{Y}$=24.775 ;$\bar{X_1}$=105; $\bar{X_2}$=103; $\bar{X_3}$=13.78	$\sum Y = n_0 b_0 + b_1 \sum X_1 + b_2 \sum X_2 + b_3 \sum X_3$ $\sum X_1 Y = b_0 \sum X_1 + b_1 \sum X_1^2 + b_2 \sum X_1 X_2 + b_3 \sum X_1 X_3$

Standard error $S_{y/x1,x2,x3} = 1.333$	$\sum X_2 Y = b_0 \sum X_2 + b_1 \sum X_1 X_2 + b_2 \sum X_2^2 + b_3 \sum X_2 X_3$ $\sum X_3 Y = b_0 \sum X_3 + b_1 \sum X_1 X_3 + b_2 \sum X_2 X_3 + b_3 \sum X_3$ $148.65 = 6b_0 + 630b_1 + 620b_2 + 82.672b_3$ $20484 = 630b_0 + 81900b_1 + 64380b_2 + 8436.39b_3$
Coefficient of determination $R^2 = 0.997$	$15153.30 = 620b_0 + 64380b_1 + 64112b_2 + 8554.955b_3$ $1973.222 = 82.672b_0 + 8436.39b_1 + 8554.955b_2 + 1143.744b_3$ where $b_0 = -140.935$; $b_1 = 0.359$; $b_2 = 1.349$; $b_3 = -0.826$.
Coefficient of correlation $r = 0.998$	And the regression equation has view of $\hat{Y}_c = -140.935 + 0.359X_1 + 1.349X_2 - 0.826X_3$ or $\sum \delta = -140.935 + 0.359N + 1.349n - 0.826V$

The statistical parameters of regression equation $\sum\delta = -140.935 + 0.359N + 1.349 n - 0.826V$:

1. Coefficient of determination $R^2 = [\sum(Y - \bar{Y})^2 - \sum(\hat{Y} - Y_c)^2] / \sum(Y - \bar{Y})^2$, where $R = 0.997$;

2. Coefficient of correlation $r = [\sum(Y - \bar{Y}) \cdot (\hat{Y}_c - \bar{Y})] / [\sum(Y - \bar{Y})^2 \cdot \sum(\hat{Y}_c - \bar{Y})^2]^{0.50}$, where $r = 0.998$;

3. Standard error $S_{y/x1,x2,x3} = \sum(Y - \hat{Y}_c)^2 / n - 3$; $S_{y/x1,x2,x3} = 1.778$;

4. Regression sum of squares $SSR = (\sum Y - \bar{Y})^2 - (\sum Y - \hat{Y}_c)^2$, where $SSR = 1529.808$.

In Figure 1.1 is shown the scatter plots of duration in-service (N), revolution per minute of engine(n),ship speed (V) versus of cumulative fouling ($\sum\delta$) regarding of equation $\sum\delta = -140.935 + 0.359N + 1.349n - 0.826V$.

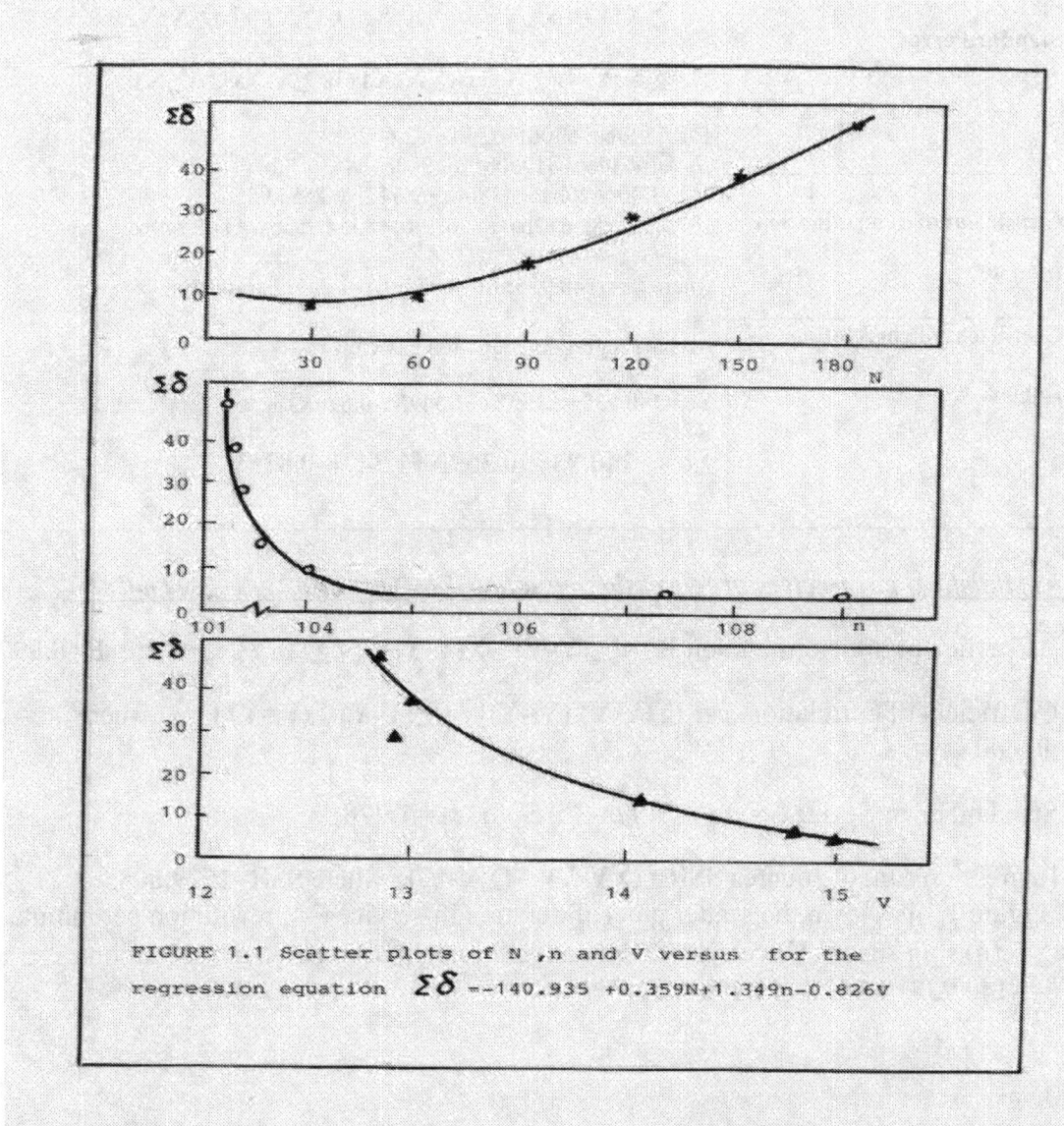

FIGURE 1.1 Scatter plots of N ,n and V versus for the regression equation $\Sigma\delta$ --140.935 +0.359N+1.349n-0.826V

In each functional view of dependency $\Sigma\delta=\gamma_1(N)$; $\Sigma\delta=\gamma_2(n)$ and $\Sigma\delta=\gamma_3(V)$ we see that regression model has the nonlinear function :

- At the first function $\Sigma\delta=\gamma_1(N)$,we see that value of fouling increases with increasing of duration in-service;
- At the second function $\Sigma\delta=\gamma_2(n)$,we see that value of fouling decreases with increasing of revolution per minute of engine;
- At the third function $\Sigma\delta=\gamma_3(V)$,we see that value of fouling decreases with increasing of ship speed.

So, from Figure 1.1 we see the main conclusions of this functional model $\Sigma\delta=\phi(N,n,V)$ that the ship in the tropics should to run faster and decrease the time of standing in ports for load and unload operations.

The better conditions for ship in the tropics is the period when this ship runs only with the high ship speed.

130

And besides in Figure and Table 2 we analyzed the functional model of changing the value of fouling in view of $\sum\delta=\phi(N,n,V)$. However, the data of Figure 1.1 shows that the big interest for the analysis have the functional model view of $\sum\delta=\gamma_1(N); \sum\delta=\gamma_2(n)$ and $\sum\delta=\gamma_3(V)$ and calculations of these models are introduced in Table 3, 4 and 5 with the different regression equations.

$^{1.404}$

Table 3 Calculation and evaluation of nonlinear regression equation $\sum\delta=0.03N$

Y	X	logX	logY	(logX)(logY)	$(logX)^2$	$(logY)^2$	$\hat{(Y_c)}$	$(Y-\bar{Y})^2$	$(Y-Y_c)^{\wedge 2}$	$(logY_c)^{\wedge 2}$	$(X-\bar{X})^2$
4.70	30	1.477	0.672	0.993	2.182	0.452	3.56	403.0	1.299	0.304	5625
9.72	60	1.778	0.988	1.757	3.162	0.975	9.41	226.7	0.096	0.948	2025
16.95	90	1.954	1.229	2.401	3.819	1.511	16.63	61.23	0.102	1.491	225
28.44	120	2.079	1.454	3.023	4.323	2.114	24.91	13.43	12.46	1.949	225
38.99	150	2.176	1.591	3.462	4.735	2.531	34.07	202.07	24.21	2.348	2025
49.85	180	2.255	1.698	3.829	5.086	2.882	44.01	628.76	34.11	2.701	5625
Total: 148.65	630	11.719	7.632	15.465	23.31	10.47		1535.1	72.27	9.741	15750

Mean: $\bar{Y}=24.775; \bar{X}=105$	Solving two normal equations we can determine the coefficients b and **loga**, and also the regression equations:
Standard error $\sigma=0.347$	$\sum logY=n_0 loga+b\sum logX$
Coefficient of determination $R^2=0.952$	$\sum(logX)(logY)= loga\sum logX+b\sum(logX)^2$ $7.632=6loga+11.719b$
Coefficient of correlation r=0.992	$15.465=11.719loga+23.307b$ where **b=1.404** ;**loga=-1.472** and regression equation $logY_c=loga+blogX; \quad \hat{Y_c}=0.03X^{1.404}$ or $\sum\delta=0.03N^{1.404}$ (4)

Statistical parameters of regression equation $\sum\delta=0.03N^{1.404}$:

1. Standard error $\sigma = [\sum(logY)^2 -\sum(logY_c)^{\wedge 2}]/n_0$, where $\sigma=0.347$;

2. Coefficient of determination $R^2 = (SST-SSE)/SST,$ where $SST=\sum(Y-\bar{Y})^2 =1535.143$, $SSE=\sum(Y-Y_c)^{\wedge 2} =72.27$ and R=0.952;

3. Coefficient of correlation $r= [\sum(X_i-\bar{X})(Y_i-\bar{Y})]/\{[\sum(X_i-\bar{X})^2]^{0.5}\cdot[\sum(Y_i-\bar{Y})^2]^{0.5}\}$, where r=0.992.

Table 4 Exponentially smoothed fouling moving average

Duration in-service ,days t	Actual series ,mm X_t	Smoothed series	
		$\alpha=0.90$ $F_t(1)$	$\alpha=0.10$ $F_t(1)$
30	4.70	4.70	4.70
60	9.72	9.22	5.20
90	16.95	16.18	6.38
120	28.44	27.21	8.59
150	38.99	37.81	11.63
180	49.85	48.65	15.45

As we from the Table 4 the better forecasting of fouling gives the result at smoothing constant $\alpha=0.90$ and for this reason should to evaluate the regression equation for the forecasting of fouling on the ship's hull.

In Table 5 is given the calculation and evaluation of forecasting regression equation $\sum \delta_F=0.05N^{1.32}$ for fouling with $\alpha=0.90$.

Table 5 Calculation and evaluation of forecasting regression equation $\sum \delta_F=0.05N^{1.32}$ for fouling with smoothing constant $\alpha=0.90$

Y	X	logX	logY	(logX)(logY)	$(logX)^2$	$(logY)^2$	$\hat{Y}_c$	$(Y-\bar{Y})^2$	$(Y-\hat{Y}_c)^2$	$(log\hat{Y}_c)^2$

4.70	30	1.477	0.672	0.993	2.182	0.452	4.61	370.95	0.008	0.440
9.22	60	1.778	0.965	1.716	3.162	0.931	11.12	217.27	3.61	1.094
16.18	90	1.954	1.209	2.362	3.819	1.462	18.99	60.53	7.89	1.635
27.21	120	2.079	1.435	2.983	4.323	2.058	27.76	10.56	0.303	2.083
37.81	150	2.176	1.578	3.434	4.735	2.489	37.27	191.82	0.292	2.469
48.65	180	2.255	1.687	3.804	5.086	2.846	47.42	609.59	1.52	2.809
Total: 143.77	630	11.719	7.546	15.292	23.307	10.24		1460.72	13.623	10.530

Mean: $\overline{Y}$=23.96 ; $\overline{X}$=105	The normal two equations for determining of coefficients **b** and **log a :**
Standard error σ=1.01	$$\sum \log Y = n \log a + b \sum \log X$$ $$\sum(\log X \log Y) = \log a \sum \log X + b \sum (\log X)^2$$ 7.546=6loga+11.719b 15.292=11.719loga+23.307b
Coefficient of determination $R^2 = 0.991$	So, the coefficients **b=1.32** and **log=-1.295** and regression equation has view of logYc=-1.295+1.32logX where
Coefficient of correlation r=0.995	$$\hat{Y}c=0.05X^{1.32} \quad \text{or} \quad \sum \delta_F = 0.05N^{1.32} \quad (5)$$

Statistical parameters of regression equation $\sum \delta_F = 0.05N^{1.32}$ **:**

1. Standard error $\sigma = (\sum \log Y_c)^2 / \sum(\log Y)^2$,where σ =1.01;

2. Coefficient of determination $R^2 = [\sum(Y-\overline{Y})^2 - \sum Y-\hat{Y}c)^2] / [\sum(Y-\overline{Y})^2]$,where R^2=0.991;

3. Coefficient of correlation r= $(R^2)^{0.5}$,where r=0.995.

7.2 Statistical nonlinear regression models for evaluation of fouling ship's hull in the tropics and ways of its prediction.

a) *Regression model function of $\sum\delta_1=\gamma_1(N)$*

As we can see from Figure 1.1 that functional dependence of fouling from duration in-service (N) ,revolution per minute of engine (n) and ship speed (V) has the nonlinear regression model.

The calculation was given in Table 3 for the regression model of function $\sum\delta_1=\gamma_1(N)$ and also the statistical characteristics of regression equation view of $\sum\delta=0.03N^{1.404}$.

In Figure 1.2 is shown the regression model of fouling ship's hull on duration in-service with the single exponentially smoothing forecasting model at the different smoothing constant ($\alpha=0.10$ and $\alpha=0.90$).

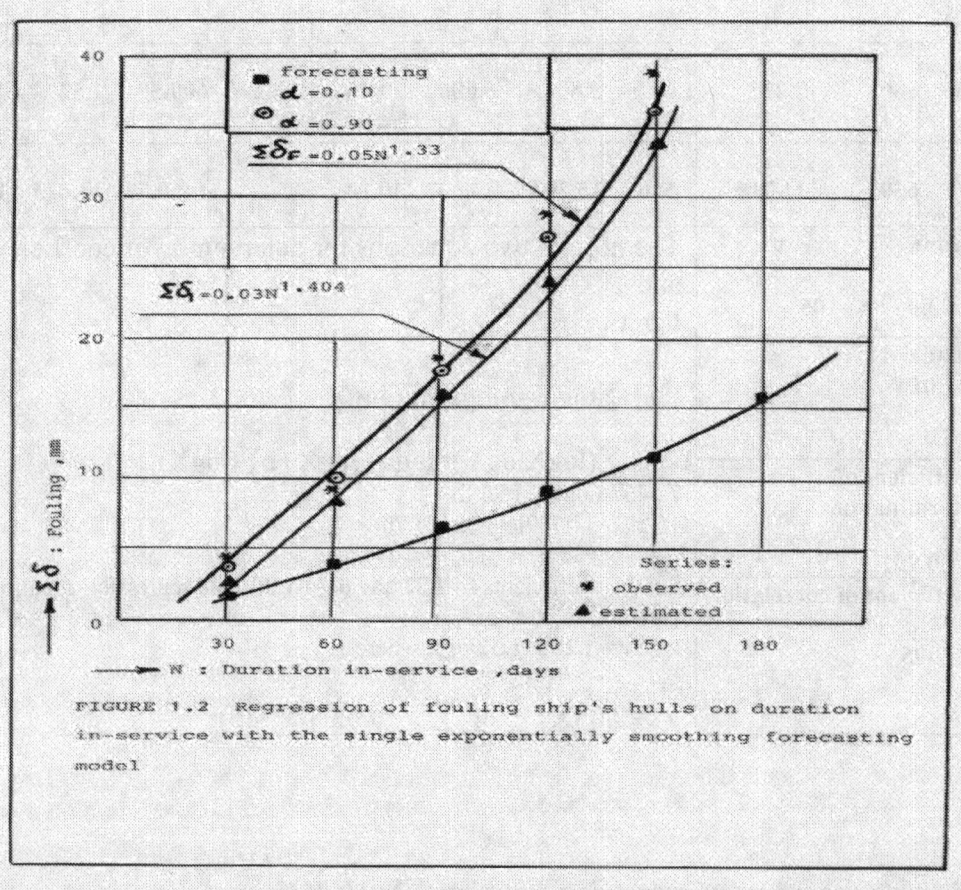

FIGURE 1.2 Regression of fouling ship's hulls on duration in-service with the single exponentially smoothing forecasting model

b) *Regression model of function $\sum\delta_2=\gamma_2(n)$:*

In Table 6 is given the calculation and evaluation of nonlinear regression equation of view

$$\sum\delta_2=2.7\cdot10^{23}\cdot n^{-11.0} \quad (6)$$

Table 6 Calculation and evaluation of nonlinear regression equation $\sum\delta_2=2.7\times10^{23-11}n$

Y	X	logX	logY	(logX)(logY)	$(logX)^2$	$(logY)^2$	$\hat{Y}_c$	$(Y-Yc)^{\wedge2}$	$(logYc)^{\wedge2}$
4.70	109	2.037	0.672	1.369	4.151	0.452	10.48	33.41	1.04
9.72	104	2.017	0.988	1.993	4.068	0.975	17.55	61.31	1.53
16.95	103	2.013	1.229	2.474	4.052	1.511	19.50	6.51	1.66
28.44	101	2.004	1.454	2.914	4.017	2.114	24.19	18.06	1.90
38.99	102	2.009	1.591	3.196	4.034	2.531	21.71	298.59	1.79
49.85	101	2.004	1.698	3.403	4.017	2.882	24.19	658.44	1.90
Total: 148.65	620	12.084	7.632	15.349	24.339	10.465		1076.3	9.82

Mean : $\bar{Y}=24.775$; $\bar{X}=103.333$	The normal two equations for determining of coefficients **b** and **loga** : $\sum logY = n_0 loga + b\sum logX$ $\sum(logX logY) = loga\sum logX + b\sum(logX)^2$ $7.632 = 6loga + 12.084b$ $15.349 = 12.084loga + 24.339b$
Standard error $\sigma=0.33$	
Coefficient of determination $R^2 =0.14$	So, the coefficients **b=-11.0** and **loga=23.43** and the regression equation has view of
Coefficient of correlation r=0.37	$\hat{Y}c=2.70\times10^{23}\times X^{-11.00}$ or $\sum\delta_2=2.7\cdot10^{23}\cdot n^{-11.0}$

Statistical parameters of regression equation $\sum\delta_2=2.7\cdot10^{23}\cdot n^{-11}$:

1. Standard error $\sigma = [\sum(logY)^2 - \sum(logYc)^{\wedge2}] / n_0$,where $\sigma=0.33$;

2.Coefficient of determination $R^2 = [\sum(logYc)^2 - (\overline{logY})\sum logY] /[\sum(logY)^2 - (\overline{logY})\sum logY]$,

Anatoly Rozenblat

where **R =0.14;**

3.Coefficient of correlation **r= (R^2)$^{0.50}$** ,where **r=0.37.**

Analysis of statistical parameters shows that coefficient of determination R^2 =0.14 (only 14 percentage) does not have a very strong nonlinear relationship between fouling and revolution per minute of engine.

In Figure 1.3 is shown the regression of fouling ship's hull on the revolution per minute of engine.

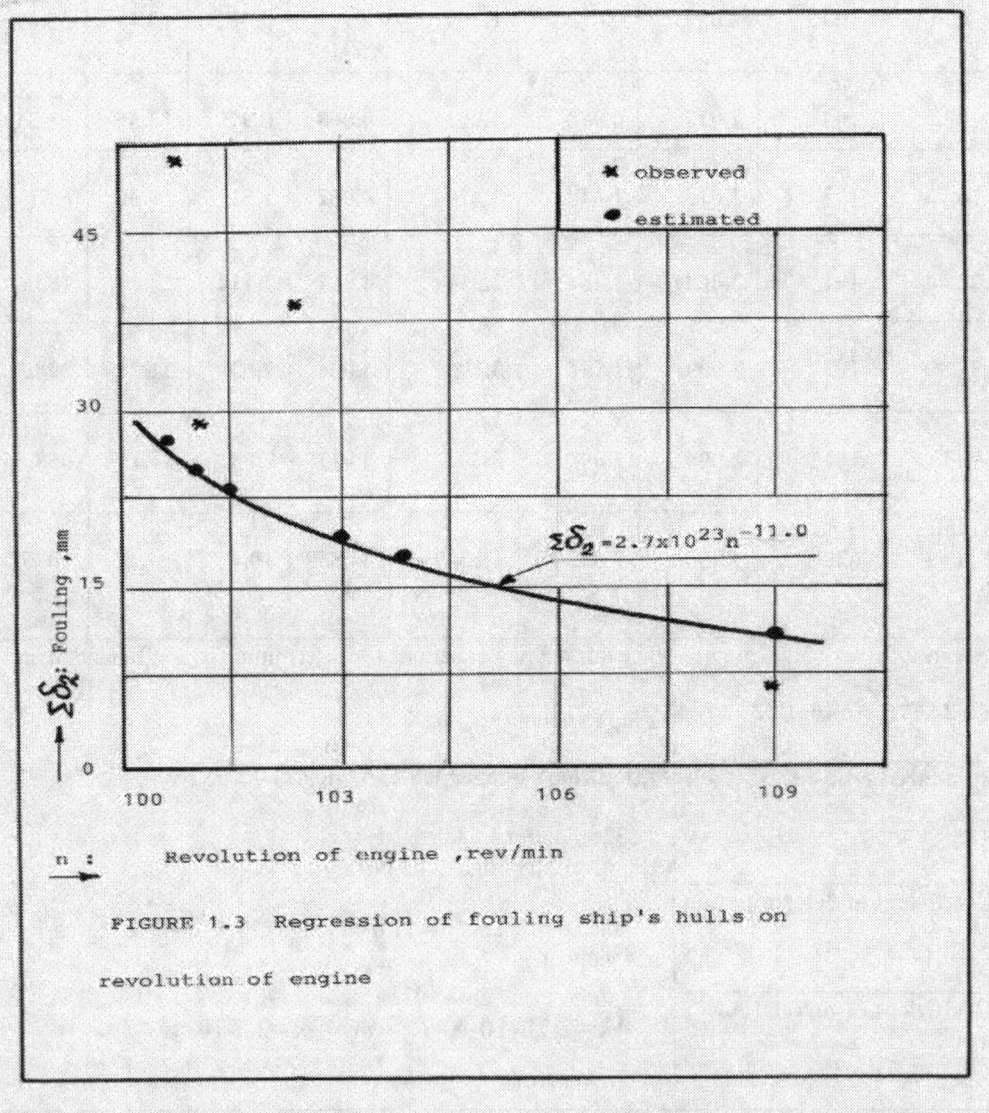

FIGURE 1.3 Regression of fouling ship's hulls on revolution of engine

As we see from Figure 1.3 this regression model has the nonlinear relationship with regression equation $\sum \delta_2 = 2.7 \cdot 10^{23} \cdot n^{-11.0}$. And besides in Figure 1.3 is shown that with decreasing of revolution per minute of engine ,the value of fouling ship's hull considerably increases.

a)   *Regression model of function $\sum \delta_3 = \gamma_\beta(V)$:*

The evaluation and calculation of this nonlinear regression model function is given in Table 7.

Table 7 Calculation and evaluation of nonlinear regression equation $\sum \delta_3 = 1.648 \cdot 10^{10} \cdot V^{-7.86}$

Y	X	logX	logY	(logX)(logY)	$(logX)^2$	$(logY)^2$	$\hat{Y}c$	$(Y-Yc)^2$	$(log\hat{Y}c)$
4.70	14.916	1.174	0.672	0.789	1.377	0.452	9.82	26.214	0.984
9.72	14.814	1.171	0.988	1.157	1.370	0.975	10.36	0.409	1.030
16.95	14.120	1.149	1.229	1.412	1.322	1.511	15.11	3.386	1.391
28.44	12.775	1.106	1.454	1.608	1.224	2.114	33.12	21.902	2.311
38.99	13.073	1.116	1.591	1.776	1.246	2.531	27.69	127.69	2.080
49.85	12.974	1.113	1.698	1.889	1.239	2.882	29.38	419.021	2.155
Total: 148.65	82.672	6.829	7.632	8.631	7.778	10.465		598.622	9.951

Mean: $\overline{Y}=24.775$; $\overline{X}=13.779$	The normal two equations for determining of coefficients **b** and **loga** have the following view: $\sum logY = n\,loga + b\sum logX$
Standard error $\sigma=0.292$	$\sum(logX\,logY) = loga\sum logX = b\sum(logX)^2$
Coefficient of determination $R^2 =0.32$	$7.632 = 6\,loga + 6.829b$ $8.631 = 6.829\,loga + 7.778b$ So, the coefficients **b=-7.76** and **loga=10.217** and regression equation h view of
Coefficient of correlation r= 0.57	$\hat{Y}c = 1.648 \cdot 10^{10} \cdot X^{-7.86}$ or $\sum\delta_3 = 1.648 \times 10^{10} \times V^{-7.86}$ (7)

And scatter plot of ship speed (V) versus of cumulative fouling $\sum\delta_3$ is illustrated in Figure 1.4.

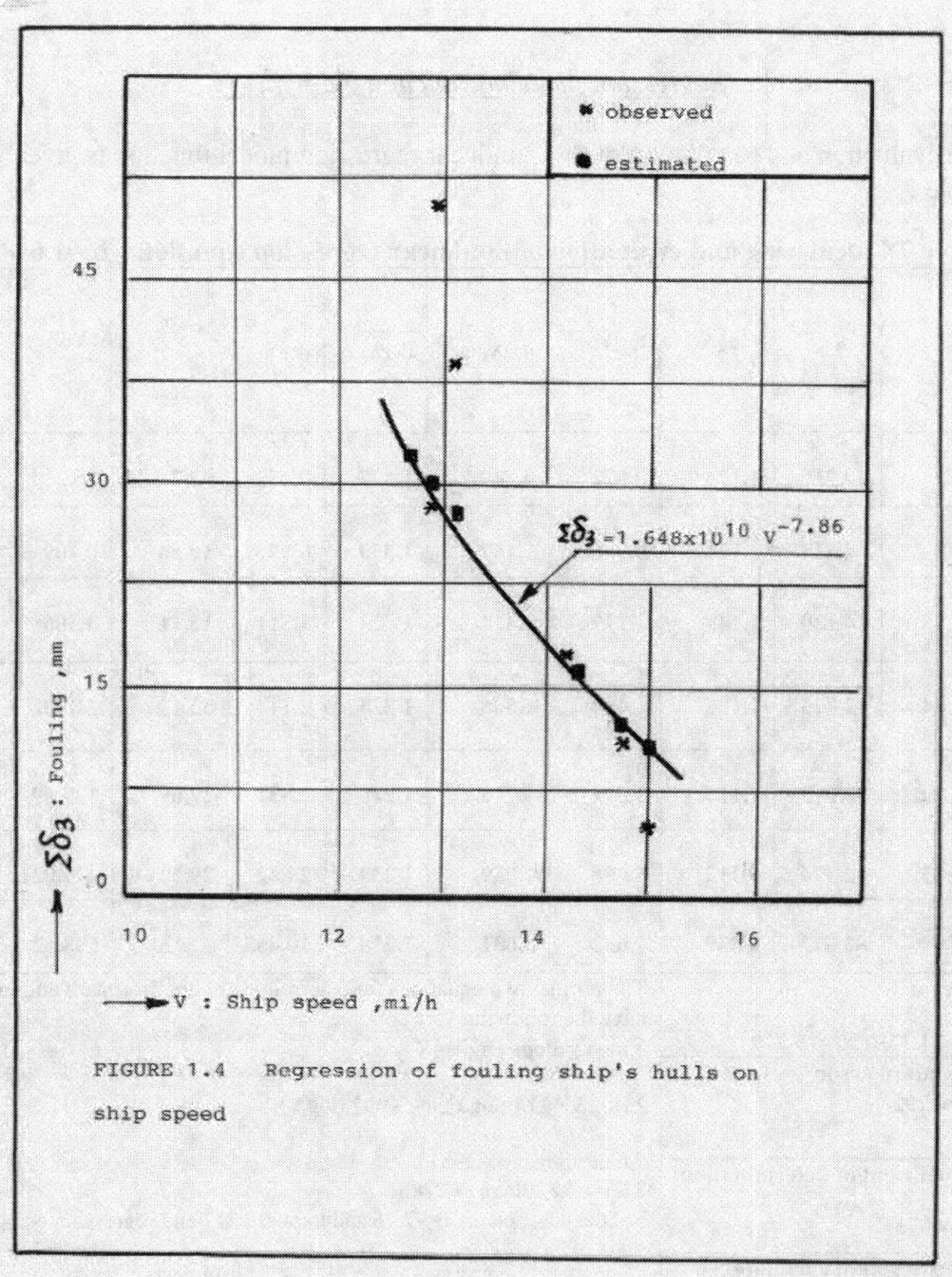

FIGURE 1.4   Regression of fouling ship's hulls on ship speed

From Figure 1.4 we see that the cumulative fouling of ship hull has the nonlinear regression model and its value increases with decreasing of ship speed.

So, the author again underlines that the best conditions for any ship in the tropical seawaters and Ocean is the way to run faster with the high ship speed, without of stopping for a long time in ports for loading and unloading operations.

10 -7.86

<u>**Statistical parameters of regression equation**</u> $\sum \delta_3 = 1.648_x 10_x V$:

1.Standard error $\sigma = [\sum (\log Y)^2 - \sum (\log Yc)^2]/n$,where $\sigma = 0.292$;

2.Coefficient of determination $R^2 = [\sum(\log Yc)^2 - (\log\hat{Y})^2\sum\log Y]/[\sum(\log Y)^2 - (\log\bar{Y})^2\sum\log Y]$,

where $R^2 = 0.32$;

3.Coefficient of correlation $r = (R^2)^{0.5}$,where $r = 0.57$.

7.3 Some anti-fouling methods and arrangements in using for ships, running in the tropics*(on the base of U.S Patents research).*

As was shown in above-named research papers ,the problem of fouling on ship's hulls is the big problem which appears in process of navigation ship in the tropics. And for this reason today many researchers work to this direction that to decrease the fouling on the ships.

So, the author **(Alec William,1973)** at his book discovers many U.S Patents which determine the general coating materials for the ship's hulls with the objective of decreasing of fouling in the process of running ship in the different seawaters and Oceans.

And besides, the author of this material additionally introduces some list **(Appendix 3)** U.S Patents for the further consideration and analysis for many researchers who have the big interest to this important problem as fouling of ship's hulls, running in the tropics.

REFERENCES

[1] Alec Williams.*Anti-fouling marine coating,1973.*
- Noyes Data Corporation

[2] U.S Patent #3493324 .*Process for protecting ship's hulls from fouling.*

[3] U.S Patent # 4098925 . *Method for protecting ship's against fouling.*

BIBLIOGRAPHY

- **Roger C. Pfaffenberger, James H.Patterson.** *Statistical methods for business and economics,1977.*
  - *Richard D.Irwing,Inc.*

- **Frederick E.Croxton, Dudley J.Cowden.** *Applied General Statistics,1939.*

Anatoly Rozenblat

-*Prentice-Hall , Inc.*

* **K.J.Rawson, C.Tupper** .*Basic Ship Theory,1968.*
-*American Elsevier Publishing Company ,Inc, New York.*

- **George W.Snedecor,William G.Cochran.** *Statistical methods.* 8th *Edition,1989.*
  - *Iowa State University Press/ AMES*

- **David G.Kleinbaum, Lawrence L.Kupper.** *Applied Regression Analysis and other multivariable methods.*
-*Doxbury Press,1978.*

Appendix 1 Ship speed (Y_i) and thirteen independent variables (X_i)
for the dry cargo ship of deidveit =10,984 ton.
(Running area = Black Sea ,Indian Ocean and other tropical seas)

Obs.	X_1	X_2	X_3	X_4	X_5	X_6	X_7
20	110	20	16	5	-	-	-
21	110	21	17	7	-	-	-
22	110	22	16	3	-	-	-
23	110	23	17	2	-	-	-
24	110	24	18	5	-	30	-
25	110	25	18	4	-	-	330
26	110	26	19	3	-	-	330
27	108	27	20	7	-	-	330
28	107	28	20	7	-	-	330
29	108	29	20	5	-	-	-
30	108	30	23	7	-	-	-
31	108	31	27	3	360	-	-
32	109	32	28	5	-	30	-
33	108	33	27	1	360	-	-
34	109	34	27	2	-	-	-
35	108	35	27	5	-	-	-
36	110	36	28	2	-	-	-
37	0	37	28	-	-	-	-
38	106	38	28	2	-	-	-
39	87	39	28	4	-	-	-

Appendix 1 Ship speed (Y_i) and thirteen independent variables (X_i)
for the dry cargo ship of deidveit =10,984 ton.
(Running area = Black Sea ,Indian Ocean and other tropical seas)

Obs.	X_1	X_2	X_3	X_4	X_5	X_6	X_7
40	87	40	28	1	360	-	-
41	107	41	28	1	360	-	-
42	107	42	28	1	360	-	-
43	0	43	28	-	-	-	-
44	0	44	28	-	-	-	-
45	0	45	28	-	-	-	-
46	0	46	28	-	-	-	-
47	0	47	28	-	-	-	-
48	0	48	28	-	-	-	-
49	0	49	28	-	-	-	-
50	0	50	29	-	-	-	-
51	0	51	29	-	-	-	-
52	0	52	28	-	-	-	-
53	0	53	28	-	-	-	-
54	0	54	28	-	-	-	-
55	0	55	28	-	-	-	-
56	0	56	28	-	-	-	-
57	0	57	28	-	-	-	-
58	0	58	28	-	-	-	-
59	0	59	28	-	-	-	-

Appendix 1   Ship speed (Y_i) and thirteen independent variables (X_i) for the dry cargo ship of deidveit = 10,984 ton.

(Running area = Black Sea , Indian Ocean and other tropical seas)

Obs.	X_1	X_2	X_3	X_4	X_5	X_6	X_7
60	0	60	28	-	-	-	-
61	0	61	28	-	-	-	-
62	0	62	28	-	-	-	-
63	0	63	28	-	-	-	-
64	0	64	29	-	-	-	-
65	0	65	29	-	-	-	-
66	0	66	29	-	-	-	-
67	0	67	29	-	-	-	-
68	0	68	29	-	-	-	-
69	0	69	29	-	-	-	-
70	0	70	29	-	-	-	-
71	0	71	29	-	-	-	-
72	0	72	29	-	-	-	-
73	0	73	29	-	-	-	-
74	0	74	29	-	-	-	-
75	0	75	28	-	-	-	-
76	0	76	28	-	-	-	-
77	0	77	28	-	-	-	-
78	0	78	28	-	-	-	-
79	0	79	28	-	-	-	-

Appendix 1 Ship speed (Y_i) and thirteen independent variables (X_i) for dry cargo ship of deidveit = 10,984 ton.

(Running area = Black Sea ,Indian Ocean and other tropical seas)

Obs.	X_1	X_2	X_3	X_4	X_5	X_6	X_7
80	0	80	29	-	-	-	-
81	0	81	29	-	-	-	-
82	0	82	29	-	-	-	-
83	0	83	29	-	-	-	-
84	0	84	29	-	-	-	-
85	0	85	29	-	-	-	-
86	0	86	29	-	-	-	-
87	0	87	25	-	-	-	-
88	0	88	25	-	-	-	-
89	0	89	25	-	-	-	-
90	0	90	25	-	-	-	-
91	0	91	26	-	-	-	-
92	0	92	26	-	-	-	-
93	0	93	26	-	-	-	-
94	0	94	26	-	-	-	-
95	0	95	26	-	-	-	-
96	0	96	26	-	-	-	-
97	0	97	26	-	-	-	-
98	0	98	26	-	-	-	-
99	0	99	26	-	-	-	-

Appendix 1   Ship speed (Y_i) and thirteen independent variables (X_i) for the dry cargo ship of deidveit = 10,984 ton.

Obs.	X_1	X_2	X_3	X_4	X_5	X_6	X_7
100	0	100	26	-	-	-	-
101	106	101	27	5	-	-	-
102	106	102	28	5	-	-	-
103	106	103	28	3	-	-	-
104	94	104	28	1	360	-	-
105	0	105	28	-	-	-	-
106	0	106	28	-	-	-	-
107	0	107	28	-	-	-	-
108	0	108	28	-	-	-	-
109	0	109	28	-	-	-	-
110	0	110	29	-	-	-	-
111	0	111	29	-	-	-	-
112	0	112	29	-	-	-	-
113	0	113	28	-	-	-	-
114	0	114	28	-	-	-	-
115	0	115	28	-	-	-	-
116	0	116	28	-	-	-	-
117	0	117	28	-	-	-	-
118	0	118	28	-	-	-	-
119	0	119	29	-	-	-	-

Appendix 1 Ship speed (Y_i) and thirteen independent variables (X_i)
for the dry cargo ship of deidveit = 10,984 ton.

(Running area = Black Sea , Indian Ocean and other tropical seas)

Obs.	X_1	X_2	X_3	X_4	X_5	X_6	X_7
120	0	120	29	-	-	-	-
121	0	121	29	-	-	-	-
122	95	122	29	1	360	-	-
123	0	123	30	-	-	-	-
124	0	124	28	-	-	-	-
125	0	125	28	-	-	-	-
126	0	126	29	-	-	-	-
127	0	127	29	-	-	-	-
128	0	128	30	-	-	-	-
129	94	129	29	1	360	-	-
130	104	130	26	1	360	-	-
131	0	131	28	-	-	-	-
132	0	132	28	-	-	-	-
133	0	133	27	-	-	-	-
134	0	134	27	-	-	-	-
135	104	135	28	1	360	-	-
136	102	136	28	1	360	-	-
137	104	137	28	1	360	-	-
138	103	138	27	1	360	-	-
139	0	139	28	-	-	-	-
140	0	140	28	-	-	-	-

Appendix 1   Ship speed (Y_i) and thirteen independent variables (X_i)
for the dry cargo ship of deidveit = 10,984 ton.

(Running area = Black Sea ,Indian Ocean and other tropical seas)

Obs.	X_1	X_2	X_3	X_4	X_5	X_6	X_7
141	0	141	28	-	-	-	-
142	0	142	28	-	-	-	-
143	0	143	28	-	-	-	-
144	104	144	23	1	360	-	-
145	104	145	23	1	360	-	-
146	104	146	22	1	360	-	-
147	103	147	23	1	360	-	-
148	105	148	23	1	360	-	-
149	103	149	22	1	360	-	-
150	103	150	22	1	360	-	-
151	103	151	22	1	-	30	-
152	103	152	22	1	-	-	330
153	103	153	18.5	2	-	-	-
154	103	154	19	2	-	30	-
155	103	155	19	3	-	-	-
156	103	156	19	1	360	-	-
157	103	157	19	1	360	-	-
158	103	158	19	1	360	-	-
159	103	159	24	1	360	-	-
160	103	160	25	3	-	-	330
161	103	161	26	2	-	-	330

Appendix 1 Ship speed (Y_i) and thirteen independent variables (X_i) for the dry cargo ship of deidveit = 10,984 ton.

(Running area = Black Sea ,Indian Ocean and other tropical seas)

Obs.	X_1	X_2	X_3	X_4	X_5	X_6	X_7
162	103	162	27	1	360	-	-
163	103	163	28	1	360	-	-
164	103	164	28	1	-	-	330
165	103	165	28	2	-	30	-
166	103	166	25	3	-	-	-
167	103	167	25	1	360	-	-
168	103	168	25	1	360	-	-
169	103	169	25	2	-	-	-
170	103	170	24	3	-	-	330
171	103	171	23	2	-	-	-
172	104	172	23	4	-	-	330
173	104	173	23	3	-	-	-
174	102	174	22	3	-	-	-
175	102	175	22	1	360	-	-
176	103	176	22	1	360	-	-
177	104	177	22	1	360	-	-
178	103	178	21	4	-	-	-
179	102	179	21	4	-	-	-
180	101	180	21	3	-	-	-
181	105	181	21	3	-	-	-
182	101	182	20	2	-	-	-
183	103	183	20	2	-	30	-

Appendix 1

Obs.	X_8	X_9	X_{10}	X_{11}	X_{12}	X_{13}	Y
1	-	-	-	-	-	400	14.25
2	-	-	-	-	-	400	13.97
3	-	-	-	-	-	405	10.86
4	-	-	-	-	-	400	13.58
5	180	-	-	-	-	415	17.25
6	-	-	210	-	-	420	14.71
7	-	-	-	-	-	410	13.84
8	-	-	-	-	-	410	13.42
9	-	-	-	-	-	410	12.88
10	-	-	-	-	-	410	11.14
11	-	-	-	-	-	412	12.47
12	-	-	-	-	-	412	12.84
13	-	-	-	-	-	411	12.77
14	-	-	-	-	-	411	13.73
15	-	-	-	-	-	412	13.19
16	-	-	-	-	270	410	12.47
17	-	-	-	-	-	410	12.60
18	-	-	-	-	-	415	13.58
19	-	-	210	-	-	408	12.69
20	-	-	-	90	-	401	12.8
21	-	150	-	-	-	407	13.23
22	180	-	-	-	-	405	13.30

Appendix 1

Obs.	X_8	X_9	X_{10}	X_{11}	X_{12}	X_{13}	Y
23	-	-	-	90	-	409	10.97
24	-	-	-	-	-	408	11.86
25	-	-	-	-	-	408	12.64
26	-	-	-	-	-	412	12.90
27	-	-	-	-	-	405	12.43
28	-	-	-	-	-	405	11.25
29	-	-	-	-	270	410	12.56
30	-	-	-	-	270	410	12.67
31	-	-	-	-	-	408	12.58
32	-	-	-	-	-	403	11.17
33	-	-	-	-	-	400	11.23
34	180	-	-	-	-	417	13.60
35	-	-	210	-	-	412	11.82
36	180	-	-	-	-	418	13.17
37	-	-	-	-	-	0	0
38	-	-	210	-	-	395	13.99
39	-	150	-	-	-	300	13.99
40	-	-	-	-	-	300	11.67
41	-	-	-	-	-	412	14.38
42	-	-	-	-	-	408	14.01
43	-	-	-	-	-	0	0
44	-	-	-	-	-	0	0

Appendix 1

Obs.	X_8	X_9	X_{10}	X_{11}	X_{12}	X_{13}	Y
45	-	-	-	-	-	0	0
46	-	-	-	-	-	0	0
47	-	-	-	-	-	0	0
48	-	-	-	-	-	0	0
49	-	-	-	-		0	0
50	-	-	-	-	-	0	0
51	-	-	-	-	-	0	0
52	-	-	-	-		0	0
53	-	-	-	-	-	0	0
54	-	-	-	-	-	0	0
55	-	-	-	-	-	0	0
56	-	-	-	-	-	0	0
57	-	-	-	-	-	0	0
58	-	-	-	-	-	0	0
59	-	-	-	-	-	0	0
60	-	-	-	-	-	0	0
61	-	-	-	-	-	0	0
62	-	-	-	-	-	0	0
63	-	-	-	-	-	0	0
64	-	-	-	-	-	0	0
65		-	-	-	-	0	0
66	-	-	-	-	-	0	0

Appendix 1

Obs.	X_8	X_9	X_{10}	X_{11}	X_{12}	X_{13}	Y
67	-	-	-	-	-	0	0
68	-	-	-	-	-	0	0
69	-	-	-	-	-	0	0
70	-	-	-	-	-	0	0
70	-	-	-	-	-	0	0
71	-	-	-	-	-	0	0
72	-	-	-	-	-	0	0
73	-	-	-	-	-	0	0
74	-	-	-	-	-	0	0
75	-	-	-	-	-	0	0
76	-	-	-	-	-	0	0
77	-	-	-	-	-	0	0
78	-	-	-	-	-	0	0
79	-	-	-	-	-	0	0
80	-	-	-	-	-	0	0
81	-	-	-	-	-	0	0
82	-	-	-	-	-	0	0
83	-	-	-	-	-	0	0
84	-	-	-	-	-	0	0
85	-	-	-	-	-	0	0
86	-	-	-	-	-	0	0
87	-	-	-	-	-	0	0

Appendix 1

Obs.	X_8	X_9	X_{10}	X_{11}	X_{12}	X_{13}	Y
88	-	-	-	-	-	0	0
89	-	-	-	-	-	0	0
90	-	-	-	-	-	0	0
91	-	-	-	-	-	0	0
92	-	-	-	-	-	0	0
93	-	-	-	-	-	0	0
94	-	-	-	-	-	0	0
95	-	-	-	-	-	0	0
96	-	-	-	-	-	0	0
97	-	-	-	-	-	0	0
98	-	-	-	-	-	0	0
99	-	-	-	-	-	0	0
100	-	-	-	-	-	0	0
101	-	-	-	90	-	425	12.6
102	-	150	-	-	-	419	11.62
103	180	-	-	-	-	425	9.75
104	-	-	-	-	-	335	10.43
105	-	-	-	-	-	0	0
106	-	-	-	-	-	0	0
107	-	-	-	-	-	0	0
108	-	-	-	-	-	0	0
109	-	-	-	-	-	0	0

Appendix 1

Obs.	X_8	X_9	X_{10}	X_{11}	X_{12}	X_{13}	Y
110	-	-	-	-	-	0	0
111	-	-	-	-	-	0	0
112	-	-	-	-	-	0	0
113	-	-	-	-	-	0	0
114	-	-	-	-	-	0	0
115	-	-	-	-	-	0	0
116	-	-	-	-	-	0	0
117	-	-	-	-	-	0	0
118	-	-	-	-	-	0	0
119	-	-	-	-	-	0	0
120	-	-	-	-	-	0	0
121	-	-	-	-	-	0	0
122	-	-	-	-	-	345	11.95
123	-	-	-	-	-	0	0
124	-	-	-	-	-	0	0
125	-	-	-	-	-	0	0
126	-	-	-	-	-	0	0
127	-	-	-	-	-	0	0
128	-	-	-	-	-	0	0
129	-	-	-	-	-	375	11.87
130	-	-	-	-	-	410	11.32
131	-	-	-	-	-	0	0

Appendix 1

Obs.	X_8	X_9	X_{10}	X_{11}	X_{12}	X_{13}	Y
132	-	-	-	-	-	0	0
133	-	-	-	-	-	0	0
134	-	-	-	-	-	0	0
135	-	-	-	-	-	410	11.56
136	-	-	-	-	-	430	11.51
137	-	-	-	-	-	425	10.56
138	-	-	-	-	-	425	10.23
139	-	-	-	-	-	0	0
140	-	-	-	-	-	0	0
141	-	-	-	-	-	0	0
142	-	-	-	-	-	0	0
143	-	-	-	-	-	0	0
144	-	-	-	-	-	415	11.28
145	-	-	-	-	-	430	11.60
146	-	-	-	-	-	420	11.41
147	-	-	-	-	-	420	11.60
148	-	-	-	-	-	420	11.60
149	-	-	-	-	-	426	11.32
150	-	-	-	-	-	426	11.32
151	-	-	-	-	-	426	11.32
152	-	-	-	-	-	426	11.32
153	-	-	-	-	270	410	11.49

Appendix 1

Obs.	X_8	X_9	X_{10}	X_{11}	X_{12}	X_{13}	Y
154	-	-	-	-	-	410	10.97
155	-	-	-	-	270	415	10.97
156	-	-	-	-	-	415	10.97
157	-	-	-	-	-	415	10.97
158	-	-	--	-	-	420	10.97
159	-	-	-	-	-	420	10.97
160	-	-	-	-	-	425	11.42
161	-	-	-	-	-	430	11.38
162	-	-	-	-	-	430	11.34
163	-	-	-	-	-	430	11.38
164	-	-	-	-	-	425	11.38
165	-	-	-	-	-	425	11.42
166	-	-	-	-	270	430	11.42
167	-	-	-	-	-	430	11.38
168	-	-	-	-	-	425	11.43
169	180	-	-	-	-	430	11.38
170	-	-	-	-	-	425	11.34
171	-	-	-	90	-	430	11.38
172	-	-	-	-	-	425	11.43
173	-	-	210	-	-	430	11.38
174	-	150	-	-	-	425	11.38
175	-	-	-	-	-	425	11.38

Appendix 1

Obs.	X_8	X_9	X_{10}	X_{11}	X_{12}	X_{13}	Y
176	-	-	-	-	-	425	11.21
177	-	-	-	-	-	430	11.21
178	-	-	-	-	270	430	11.21
179	-	-	-	-	270	430	11.21
180	-	-	-	90	-	430	11.21
181	-	-	-	90	-	430	11.12
182	-	-	-	-	270	430	11.12
183	-	-	-	-	-	430	11.12

Appendix 2   Analysis of empirical formula for evaluation of ship speed in the tropics and wave sea.

Obs. (n)	Ship speed observed (Y)	Ship speed evaluated ($\hat{Y}$)	(Y - $\hat{Y}$)	(Y - $\hat{Y}$)^2	Y^2	$\hat{Y}^2$
1	14.25	13.531	0.719	0.517	203.063	183.052
2	13.97	14.845	-0.875	0.766	195.161	220.374
3	10.86	15.241	-4.381	19.193	117.939	232.288
4	13.58	13.505	0.075	0.006	184.416	182.385
5	17.25	14.607	2.643	6.985	297.563	213.364
6	14.71	14.87	-0.16	0.026	216.384	221.117
7	13.84	13.377	0.463	0.214	191.546	178.944
8	13.42	13.368	0.052	0.003	180.096	178.703
9	12.88	17.114	-4.234	17.927	165.894	292.889
10	11.14	13.844	-2.704	7.312	124.099	191.656
11	12.47	15.297	-2.827	7.992	155.501	233.998
12	12.84	15.288	-2.448	5.993	164.866	233.723
13	12.77	14.796	-2.026	4.105	163.073	218.922
14	13.73	13.082	0.648	0.419	188.513	171.139
15	13.19	14.051	-0.861	0.741	173.976	197.431
16	12.47	12.079	0.391	0.153	155.501	145.902
17	12.60	14.851	-2.251	5.067	158.76	220.552
18	13.58	13.23	0.35	0.123	184.416	175.033
19	12.69	13.89	-1.20	1.44	161.036	192.932
20	12.80	5.12	7.68	58.98	163.84	26.214

Appendix 2

(n)	(Y)	($\hat{Y}$)	(Y - $\hat{Y}$)	(Y - $\hat{Y}$)^2	Y^2	$\hat{Y}^2$
21	13.23	15.215	- 1.985	3.94	175.033	231.496
22	13.30	14.066	-0.766	0.587	176.89	197.852
23	10.97	3.536	7.434	55.264	120.341	12.503
24	11.86	15.00	-3.14	9.859	140.659	225.00
25	12.64	15.512	-2.872	8.248	159.769	240.622
26	12.90	14.968	-2.068	4.277	166.41	224.041
27	12.43	16.726	-4.296	18.456	154.505	279.759
28	11.25	16.577	-5.327	28.377	126.563	274.797
29	12.56	10.695	1.865	3.478	157.754	114.383
30	12.67	11.675	0.995	0.990	160.529	136.306
31	12.58	13.893	-1.313	1.724	158.256	193.015
32	11.17	14.84	-3.67	13.469	124.769	220.226
33	11.23	12.969	-1.739	3.024	126.113	222.786
34	13.60	13.204	0.396	0.157	184.96	174.346
35	11.82	14.414	-2.594	6.729	139.712	207.763
36	13.17	13.316	-0.146	0.021	173.449	177.316
37	0	0.851	-0.851	0.724	0	0.724
38	13.99	12.798	1.192	1.421	195.720	163.789
39	13.99	11.434	2.556	6.533	195.720	130.736
40	11.67	9.977	1.693	2.866	136.189	99.541
41	14.38	11.636	2.744	7.529	206.764	135.397
42	14.01	11.668	2.342	5.485	196.280	136.142
43	0	0.798	-0.798	0.637	0	0.637

Appendix 2

(n)	(Y)	($\hat{Y}$)	(Y - $\hat{Y}$)	(Y - $\hat{Y}$)^2	Y^2	$\hat{Y}^2$
44	0	0.789	-0.789	0.623	0	0.623
45	0	0.78	-0.78	0.608	0	0.608
46	0	0.772	-0.772	0.596	0	0.596
47	0	0.763	-0.763	0.582	0	0.582
48	0	0.754	-0.754	0.569	0	0.569
49	0	0.746	-0.746	0.557	0	0.557
50	0	0.736	-0.736	0.542	0	0.542
51	0	0.727	-0.727	0.529	0	0.529
52	0	0.719	-0.719	0.517	0	0.517
53	0	0.709	-0.709	0.503	0	0.503
54	0	0.701	-0.701	0.491	0	0.491
55	0	0.693	-0.693	0.48	0	0.48
56	0	0.684	-0.684	0.468	0	0.468
57	0	0.674	-0.674	0.454	0	0.454
58	0	0.666	-0.666	0.444	0	0.444
59	0	0.658	-0.658	0.433	0	0.433
60	0	0.649	-0.649	0.421	0	0.421
61	0	0.639	-0.639	0.408	0	0.408
62	0	0.631	-0.631	0.398	0	0.398
63	0	0.622	-0.622	0.387	0	0.387
64	0	0.613	-0.613	0.378	0	0.378
65	0	0.605	-0.605	0.366	0	0.366
66	0	0.596	-0.596	0.355	0	0.355

Appendix 2

(n)	(Y)	$(\hat{Y})$	$(Y - \hat{Y})$	$(Y - \hat{Y})^2$	Y^2	$\hat{Y}^2$
67	0	0.587	-0.587	0.345	0	0.345
68	0	0.579	-0.579	0.335	0	0.335
69	0	0.569	-0.569	0.324	0	0.324
70	0	0.56	-0.56	0.314	0	0.314
71	0	0.552	-0.552	0.305	0	0.305
72	0	0.543	-0.543	0.295	0	0.295
73	0	0.534	-0.534	0.285	0	0.285
74	0	0.526	-0.526	0.277	0	0.277
75	0	0.517	-0.517	0.267	0	0.267
76	0	0.508	-0.508	0.258	0	0.258
77	0	0.499	-0.499	0.249	0	0.249
78	0	0.491	-0.491	0.241	0	0.241
79	0	0.482	-0.482	0.232	0	0.232
80	0	0.473	-0.473	0.224	0	0.224
81	0	0.464	-0.464	0.215	0	0.215
82	0	0.455	-0.455	0.207	0	0.207
83	0	0.447	-0.447	0.199	0	0.199
84	0	0.438	-0.438	0.192	0	0.192
85	0	0.429	-0.429	0.184	0	0.184
86	0	0.421	-0.421	0.177	0	0.177
87	0	0.412	-0.412	0.169	0	0.169
88	0	0.403	-0.403	0.162	0	0.162
89	0	0.394	-0.394	0.155	0	0.155

Appendix 2

(N)	(Y)	($\hat{Y}$)	(Y-$\hat{Y}$)	(Y-$\hat{Y}$)^2	Y^2	$\hat{Y}^2$
90	0	0.386	-0.386	0.149	0	0.149
91	0	0.377	-0.377	0.142	0	0.142
92	0	0.368	-0.368	0.135	0	0.135
93	0	0.359	-0.359	0.129	0	0.129
94	0	0.35	-0.35	0.123	0	0.123
95	0	0.341	-0.341	0.116	0	0.116
96	0	0.333	-0.333	0.111	0	0.111
97	0	0.324	-0.324	0.105	0	0.105
98	0	0.315	-0.315	0.099	0	0.099
99	0	0.307	-0.307	0.094	0	0.094
100	0	0.298	-0.298	0.089	0	0.089
101	12.6	3.609	8.991	80.838	158.76	13.024
102	11.62	12.831	-1.211	1.466	135.024	164.635
103	9.75	12.589	-2.839	8.059	95.063	158.483
104	10.43	11.701	-1.271	1.615	108.785	136.913
105	0	0.254	-0.254	0.065	0	0.065
106	0	0.245	-0.245	0.060	0	0.060
107	0	0.236	-0.236	0.056	0	0.056
108	0	0.228	-0.228	0.052	0	0.052
109	0	0.219	-0.219	0.048	0	0.048
110	0	0.209	-0.209	0.044	0	0.044
111	0	0.201	-0.201	0.04	0	0.04
112	0	0.192	-0.192	0.037	0	0.037

Appendix 2

(n)	(Y)	($\hat{Y}$)	($Y-\hat{Y}$)	($Y-\hat{Y}$)^2	Y^2	$\hat{Y}^2$
113	0	0.183	-0.183	0.033	0	0.033
114	0	0.175	-0.175	0.031	0	0.031
115	0	0.166	-0.166	0.028	0	0.028
116	0	0.156	-0.156	0.024	0	0.024
117	0	0.149	-0.149	0.022	0	0.022
118	0	0.139	-0.139	0.019	0	0.019
119	0	0.130	-0.130	0.017	0	0.017
120	0	0.122	-0.122	0.015	0	0.015
121	0	0.113	-0.113	0.013	0	0.013
122	11.95	10.923	1.024	1.049	142.803	119.12
123	0	0.095	-0.095	0.009	0	0.009
124	0	0.087	-0.087	0.008	0	0.008
125	0	0.078	-0.078	0.006	0	0.006
126	0	0.069	-0.069	0.005	0	0.005
127	0	0.059	-0.059	0.003	0	0.003
128	0	0.051	-0.051	0.003	0	0.003
129	11.87	10.416	1.453	2.111	140.897	108.493
130	11.32	11.455	-0.137	0.019	128.142	131.217
131	0	0.025	-0.025	0.0006	0	0.0006
132	0	0.017	-0.017	0.0003	0	0.0003
133	0	0.0082	-0.0082	0.0001	0	0
134	0	-0.0006	0.0006	0	0	0
135	11.56	11.411	0.1450	0.021	133.634	130.211

Appendix 2

(n)	(Y)	($\hat{Y}$)	(Y-$\hat{Y}$)	(Y-$\hat{Y}$)^2	Y^2	$\hat{Y}^2$
136	11.51	10.919	0.594	0.353	132.480	119.225
137	10.56	11.241	-0.684	0.468	11.514	126.36
138	10.23	11.092	-0.861	0.741	104.653	123.033
139	0	-0.044	0.044	0.0019	0	0.0019
140	0	-0.053	0.053	0.0028	0	0.0028
141	0	-0.051	0.051	0.0026	0	0.0026
142	0	-0.071	0.071	0.005	0	0.005
143	0	-0.079	0.079	0.0062	0	0.0062
144	11.28	11.282	-0.002	0	127.238	127.284
145	1160	11.12	0.48	0.2304	134.56	123.654
146	11.41	11.214	0.196	0.0384	130.188	125.754
147	11.60	11.065	0.535	0.286	134.56	122.434
148	11.60	11.329	0.271	0.073	134.56	128.346
149	11.32	10.986	0.334	0.111	128.142	120.692
150	11.32	10.977	0.343	0.118	128.142	120.495
151	11.32	10.744	0.576	0.332	128.142	115.434
152	11.32	11.749	-0.429	0.184	128.142	138.039
153	11.49	7.423	4.067	16.54	132.020	55.190
154	10.97	11.375	-0.405	0.164	120.341	129.391
155	10.97	7.849	3.121	9.741	120.341	61.607
156	10.97	10.035	0.935	0.874	120.341	100.701
157	10.97	11.028	-0.058	0.003	120.341	121.617
158	10.97	10.968	0.002	0	120.341	120.297

Appendix 2

(n)	(Y)	($\hat{Y}$)	($Y-\hat{Y}$)	($Y-\hat{Y}$)^2	Y^2	$\hat{Y}^2$
159	10.97	10.958	0.012	0.0001	120.341	120.078
160	11.42	12.728	-1.728	2.986	130.416	162.001
161	11.38	12.123	-0.743	0.552	129.504	146.967
162	11.34	10.831	0.509	0.259	128.596	117.311
163	11.38	10.821	0.559	0.312	129.504	117.096
164	11.38	11.653	-0.273	0.075	129.504	135.792
165	11.42	11.125	0.295	0.087	130.416	123.766
166	11.42	7.600	3.82	14.592	130.416	57.760
167	11.38	10.787	0.593	0.352	129.504	116.359
168	11.42	11.323	0.097	0.009	129.504	128.210
169	11.38	11.045	0.335	0.112	129.504	121.992
170	11.34	12.589	-1.249	1.560	128.596	158.483
171	11.38	1.04	10.34	106.916	129.504	1.082
173	11.42	13.206	-1.786	3.189	129.504	174.398
173	11.38	11.472	-0.092	0.008	129.504	131.607
174	11.38	10.589	0.791	0.626	129.504	112.127
175	11.38	10.628	0.752	0.565	129.504	112.954
176	11.21	10.759	0.451	0.203	125.664	115.756
177	11.21	10.839	0.371	0.138	125.664	117.484
178	11.21	7.989	3.221	10.375	125.664	63.824
179	11.21	7.840	3.37	11.357	125.664	61.466
180	11.21	1.175	10.035	100.701	125.664	1.381
181	11.12	1.728	9.392	88.209	123.654	2.986

Appendix 2

(n)	(Y)	($\hat{Y}$)	(Y-$\hat{Y}$)	(Y-$\hat{Y}$)^2	Y^2	$\hat{Y}^2$
182	11.12	6.685	4.435	19.669	123.654	44.689
183	11.12	10.917	0.203	0.041	123.654	119.181

APPENDIX 3

6347911	Vortex shedding strake wraps for submerged pilings and pipes
6342386	Methods for removing undesired growth from a surface
6327996	Biosecure zero exchange system for maturation and growout of marine animals
6281196	Isolation and structural elucidation of the human cancer cell growth inhibitory compound denominat
6209472	Apparatus and method for inhibiting fouling of an underwater surface
6186702	Artificial reef
6173669	Apparatus and method for inhibiting fouling of an underwater surface
6156352	Method and means for husbanding marine organisms
6153251	Nutrition enriched composition for feed
6152061	Floating collapsible hull protector against marine growth
6136781	LH RH receptor antagonists
6067922	Copper protected fairings
6060429	Composition and method for controlling plant diseases caused by fungi
6044798	Floating aquaculture apparatus
6009823	Marine scoop strainer with cleaning access
6008244	Halopropargyl compounds as marine antifouling agents
5994268	Photocatalytic surfacing agents with varying oxides for inhibiting algae growth
5981435	Methods for controlling algae
5970917	Marine aquaculture apparatus
5969211	Pantropic retroviral vectors for gene transfer in mollusks
5967085	Sea water well driven heat exchange system coupled to an agricultural system and aquaculture pre
5964174	Anti fouling protective cover for stern drive unit
5961831	Automated closed recirculating aquaculture filtration system and method
5919689	Marine antifouling methods and compositions
5883120	Antifungal activity of the spongistatins
5880067	Photocatalytic surfacing agents with varying oxides for inhibiting algae growth
5855654	Pyridazinones as marine antifouling agents
5853463	Marine antifouling agent
5846936	Growth hormone releasing factor analogs
5836265	Reef ball
5833742	Phenylamides as marine antifouling agents
5814496	Process for demethylating S methyl mercapto compounds
5801222	Isolation and structure of the human cancer cell growth inhibitory cyclic octapeptides phakellistatin
5769019	Protective covering for outdoor structures
5765968	Apparatus for eliminating and preventing marine growth on offshore structures
5735226	Marine anti fouling system and method
5614006	Anti fouling composition
5587313	Method for fermentation of marine bacteria
5574057	Naamidine A extracted from sea sponges and methods for its use as an anti tumor agent
5567221	Compositions and methods for use in aquaculture
5564369	Reef ball
5549069	Enclosure for shielding moored water vessel hull from direct contact with water
5518992	Photocatalytic surfacing agents for inhibiting algae growth
5518990	Method for preventing emergence of algae and antialgal composition
5516236	Timber pile protection system
5514689	Cribrostatins 1 and 2
5491409	Multiple yoke eddy current technique for detection of surface defects on metal components covered
5465676	Barnacle shield
5431122	Apparatus for cleaning the submerged portion of ship hulls
5430053	Isolation and structure of dictyostatin 1
5421413	Flexible fairings to reduce vortex induced vibrations
5393897	Isolation and structure of spongistatins 5,7,8 and 9
5354603	Antifouling/anticorrosive composite marine structure
5351640	Portable manual boat hull cleaner
5288704	Synergistic composition comprising a fibroblast growth factor and a sulfated polysaccharide, for use
5279368	Anti fouling covering for use in sub sea structures
5279244	Combined mooring slip and underwater body protector against marine growth

APPENDIX 3

5278069	Bioleaching method for the extraction of metals from coal fly ash using thiobacillus
5231018	Extraction of metal oxides from coal fly ash by microorganisms and a new microorganism useful the
5228830	Wicket gate
5192667	Method for evaluating anti fouling paints
5146644	Pipeline cleaning device
5143011	Method and apparatus for inhibiting barnacle growth on boats
5130251	Stress resistant bioluminescent dinoflagellates
5104618	COMPOSITION FOR TREATING NETTING
5097795	WATER PURIFICATION SYSTEM AND APPARATUS
5082722	PROCESS FOR TREATING NETTING WITH AN ANTIFOULING COMPOSITION AND PRODUCT
5041713	APPARATUS AND METHOD FOR APPLYING PLASMA FLAME SPRAYED POLYMERS
5040923	APPARATUS FOR THE PREVENTING OF MARINE GROWTH OF OFFSHORE STRUCTURES
5038697	VARIABLE BEAM FLOAT CONNECTION ASSEMBLIES FOR TRIMARANS
5036785	LIGHTNING PROTECTION INSTALLATION ON A BOAT
5027550	APPARATUS FOR CULTIVATING AQUATIC LIVING THINGS IN SEA WATER
5026212	APPARATUS FOR THE COMBATTING OF MARINE GROWTH ON OFFSHORE STRUCTURES
5015372	TOXIN CONTAINING PERFORATED ANTIFOULING POLYMER NOZZLE GROMMET
5011615	METHOD AND APPARATUS FOR INHIBITING ORGANISM GROWTH IN MARINE MOTORS
5007377	APPARATUS AND METHOD FOR MARINE HABITAT DEVELOPMENT
4966096	WATER PURIFICATION SYSTEM AND APPARATUS
4923730	ANTI FOULING SURFACE STRUCTURE, ANTI FOULING COVERING MATERIAL AND METHOD
4919702	FERTILIZER AND/OR SOIL AMENDMENT
4918005	METHOD OF QUANTITATIVE ASSAY FOR VITAMIN B.SUB.12 AND REAGENT FOR ASSAYING
4911849	METHOD AND MEANS OF AERATION POWERED WATER FLOW USING FOIL SHAPED CONT
4910912	AQUACULTURE IN NONCONVECTIVE SOLAR PONDS
4909173	SCRUBBING DEVICE FOR SUBMERGED SURFACES OF BOAT HULLS AND THE LIKE
4896626	SHELLFISH CULTURE METHODS AND APPARATUS
4871551	PIGMENTATION SUPPLEMENTS FOR ANIMAL FEED COMPOSITIONS
4869013	SNAG PRUFFER
4846966	TRASH RACK
4846870	FERTILIZER AND/OR SOIL AMENDMENT
4830540	METHOD FOR CONSTRUCTING INSPECTABLE WELDED JOINTS WHICH ARE RESISTANT TO
4809381	APPARATUS FOR REMOVING MARINE GROWTH FROM PYLONS
4789005	Marine growth retarding hose
4786496	Immunopotentiating agent having anti tumor activity
4781139	One man manual boat hull cleaning device
4774345	Amine complexed zinc salts of organic diacids
4626283	CORROSION AND MARINE GROWTH INHIBITING COMPOSITIONS
4552813	METHOD OF INHIBITING THE GROWTH OF MARINE LIFE ON SURFACES IN CONTACT WITH
4415293	OFFSHORE PLATFORM FREE OF MARINE GROWTH AND METHOD OF REDUCING PLATFOR
4337716	MARINE GROWTH WIPER
4324784	PROCESS FOR PREVENTING GROWTH OF MARINE ORGANISMS ON A SUBSTANCE USING
4133862	METHOD OF INHIBITING AND/OR ERADICATING MARINE FUNGAL GROWTH WITH OBTUSAS
4076619	HYDROPHILIC ACRYLIC POLYMERS AS MARINE FILTERS, ALGAE GROWTH CATALYSTS, AN
4058075	MARINE LIFE GROWTH INHIBITOR DEVICE
4017370	METHOD FOR PREVENTION OF FOULING BY MARINE GROWTH AND CORROSION UTILIZIN
3884018	HARVESTING MARINE GROWTHS AND PACKAGING THE TREATED PRODUCT
3866396	REMOVAL OF MARINE GROWTHS FROM LAKES, WATERWAYS, AND OTHER BODIES OF WA
3709195	METHOD FOR HATCHING AND GROWING MARINE ORGANISMS
3706185	APPARATUS FOR REMOVING MARINE GROWTHS AND ROOTS
3546858	HARVESTING MARINE GROWTHS
3540194	METHOD OF REMOVING MARINE GROWTHS AND ROOTS
3443545	ARRANGEMENT FOR CLEANING OF A SHIP'S HULL OF MARINE GROWTH

APPENDIX 3

6221247	Dioxole coated membrane module for ultrafiltration or microfiltration of aqueous suspensions
6172132	Antifouling coating composition, coating film formed from said antifouling coating composition, ant
5989517	Process for producing stabilized magnesium hydroxide slurries
5904988	Sprayable, condensation curable silicone foul release coatings and articles coated therewith
5824279	Process for producing stabilized magnesium hydroxide slurries
5717007	Anti fouling self polishable paints
5603755	Preventive agent for fouling organisms
5409537	Laser coating apparatus
5354603	Antifouling/anticorrosive composite marine structure
5200230	Laser coating process
5192667	Method for evaluating anti fouling paints
5190580	Process of using biocidal compositions
5116407	Antifouling coatings
5088432	ANTI FOULING SYSTEM FOR SUBSTANCES IN CONTACT WITH SEAWATER
5080926	ANTI FOULING COATING PROCESS
5071479	BIOCIDAL COMPOSITIONS
4687792	ERODIBLE SHIP BOTTOM PAINTS FOR CONTROL OF MARINE FOULING
4594365	ERODIBLE SHIP BOTTOM PAINTS FOR CONTROL OF MARINE FOULING
4593055	ERODIBLE SHIP BOTTOM PAINTS FOR CONTROL OF MARINE FOULING
4518638	METHOD FOR THE PROTECTION OF SHIPS AND OTHER OBJECTS AGAINST FOULING
4098925	METHOD FOR PROTECTING SHIPS AGAINST FOULING
3912519	ANTI FOULING SHIP BOTTOM PAINT
3832190	ANTI FOULING PAINT FOR SHIP'S BOTTOM AND STRUCTURES UNDER SEA WATER
3817759	COATING FOR PREVENTING THE FOULING OF SHIPS' PARTS
3740192	METHOD FOR THE PREVENTION OF FOULING OF SHIPS
3650924	METHOD FOR PREVENTING FOULING OF SHIPS, PARTICULARLY SHIPS HAVING SUBSTAN
3610195	DE FOULING OF SHIP'S HULLS
3493324	PROCESS FOR PROTECTING SHIP'S HULLS FROM FOULING

6209472	Apparatus and method for inhibiting fouling of an underwater surface
5964174	Anti fouling protective cover for stern drive unit
5890835	Hydraulic lift for boats
5795216	Abrading tool having a suction system for collecting abraded particles
5787847	Oil supply system for a planing type boat
5629045	Biodegradable nosiogenic agents for control of non vertebrate pests
5441368	Anti fouling apparatus for submerged marine surfaces
5397385	Anti fouling coating composition containing capsaicin
5297363	Portable surface preparation abrading unit
5259701	Antifouling coating composition comprising furan compounds, method for protecting aquatic structu
5248221	Antifouling coating composition comprising lactone compounds, method for protecting aquatic struc
5168823	Transportable off shore boat mooring and method for using same
5035759	METHOD OF PROTECTING HULLS OF MARINE VESSELS FROM FOULING
5017090	VARIABLE PITCH PROPELLER BLADES AND DRIVE AND ADJUSTING MECHANISM THEREFO
4772344	Method of protecting the hulls of marine vessels from fouling
4282822	BOAT HULL ANTI FOULING SHROUD
4280439	BOAT HULL ANTI FOULING SHROUD
4280438	BOAT HULL ANTI FOULING SHROUD
4280437	BOAT HULL ANTI FOULING SHROUD
4280436	BOAT HULL ANTI FOULING SHROUD
4215644	BOAT HULL ANTI FOULING SHROUD
4127687	PREVENTION OF FOULING OF MARINE STRUCTURES SUCH AS BOAT HULLS

6270671	Method and apparatus for microfiltration
5423963	Fouling compensation in an oxygen analyzer
5181482	Sootblowing advisor and automation system
4822475	METHOD FOR DETERMINING THE FOULING TENDENCY OF CRUDE PETROLEUM OILS
4686853	METHOD FOR THE PREDICTION AND DETECTION OF CONDENSER FOULING

ABOUT THE AUTHOR

* *Anatoly I.Rozenblat is the Independent Scientist and Inventor.*

Russian-born Anatoly I.Rozenblat has worked as a seaman, an operational ship's mechanic and marine engineer. Using the experience he gained during his twenty years working in marine technology, Rozenblat has and compiled *Analysis of Ship Speed and Engine Parameters in the Tropics.*

Rozenblat holds a bachelor of science degree in mechanical engineering from Odessa Institute of Maritime Transport Engineering and bachelor of science degree in computer and information systems from Chicago East-West University.

Rozenblat lives in Chicago, Illinois and is the father of two children, Moshe and Inna.

He is the member of the American Society of Mechanical Engineers ,the Society of Naval Architects and Marine Engineers.

His biography has been published by the International Biographical Centre, England ; the American Biographical Institute; and Marquis ' *Who's who in America'.*

He has published more than 50 articles in the technical literature and has about of 30 innovations and 6 books.